W. HOCK HOCHHEIM'S

HAND, STICK, KNIFE, GUN
CLOSE QUARTER COMBATIVES

TRAINING MISSION FIVE

CQCG

THE HOCHHEIM GROUP

CQC GROUP: TRAINING MISSION FIVE

by W. Hock Hochheim

Also by W. Hock Hochheim

Non Fiction:
Impact Weapon Combatives
Knife Counter/Knife Combatives
Training Mission Four
Training Mission Three
Training Mission Two
Training Mission One
The Foundation: Knife Fighting Encyclopedia Volume 1
Military Knife Combat: Knife Fighting Encyclopedia Volume 2
Unarmed vs. the Knife: Knife Fighting Encyclopedia Volume 3
Shooting from the Hip
Find Missing Persons
The Great Escapes of Pancho Villa

Fiction:
My Gun is My Passport
Be Bad Now

Copyright 2006, 2012
All rights reserved

ISBN Number: 978-1-932113-52-5

WARNINGS!

This CQC Group course contains a wide spectrum of less-than-lethal and lethal training. Citizens, law enforcement, military and security personnel are expected to understand the moral, legal and ethical use of force continuum, and use the same behavior, maturity and restraint in unarmed, edged and impact weapon combatives, as they do when using firearms in their line of professional duty or in the act of defending themselves or others.

Consult with medical authorities and make certain you are in physical shape before you begin this or any active course.

TRAINING MISSION FIVE

Table of Contents

1) The CQC Group: Overview/Review   **page 9**

2) Unarmed Combatives   **page 11**
    UC Strike 5:   The Punch/Counter Punch Module
    UC Kick 5:   The Lead Leg Hook Kick Module
    UC Takedown 5:   The Frontal Takedowns Module

3) SDMS Impact Weapon Combatives   **page 76**
    SDMS Level 5:   Disarms and Counters Module

4) Knife/Counter-Knife Combatives   **page 118**
    Knife Level 5:   The Reverse Grip Stab Assault Module

5) Gun/Counter-Gun Combatives   **page 155**
    Gun Level 5:   The Long Gun Disarm and Retention Module

Acknowledgements   **page 200**

CQCG

Course Review

CQCG Training Mission Progression Overview:
All training is expressed in a *Modular Concept*. A module includes learning the basic execution of a particular tactic, troubleshooting common counters to the tactic, skill and flow drill development of the tactic, counters, and using the tactic in standing, kneeling and ground positions, fighting against unarmed, stick, knife and gun weaponry. The bold/highlighted levels appear in this book.

CQCG Unarmed Combatives

The Strike Modules

Level 1	The Finger Strike Module
Level 2	The Palm Strike Module
Level 3	The Forearm Strike Module
Level 4	The Hammer Fist Module
Level 5	**Fistfight! The Punch/Counter-Punch Strike Module**
Level 6	The Elbow Strike Module
Level 7	The Body Ram Module
Level 8	The Limited Use/Head Butt Module
Level 9	The Strike/Block/Counter-Strike Module
Level 10	The Combat Scenario Performance Module

The Kick Modules

Level 1	The Frontal Snap Kick Module
Level 2	The Stomp Kick Module
Level 3	The Knee Strike Module
Level 4	The Rear Leg Round Kick Module
Level 5	**The Front Leg Hook Kick Module**
Level 6	The Back Kick Module
Level 7	The Side Kick Module
Level 8	The Front Thrust Module
Level 9	The Counters to Kicks Module
Level 10	The Combat Scenario Performance Module

The Takedowns and Throws Modules (these include joint crank studies)

Level 1	The Finger Attack Takedowns Module
Level 2	The Circular/Wheel Takedowns Module
Level 3	The Rear Takedowns Module
Level 4	The Bent and Straight Arm Takedowns Module
Level 5	**The Front Takedowns Module**
Level 6	The Neck Attack Takedowns Module
Level 7	The Push/Pull Takedowns Module
Level 8	The Tackle Takedowns Module
Level 9	The Leg Attack Takedowns Module
Level 10	The Combat Scenario Performance Module

CQCG Knife/Counter-Knife Course
- Level 1 — Knife Introduction and Quick Draw Combat Module
- Level 2 — The Saber Grip Slash Knife Module
- Level 3 — Reverse Grip Slash Knife Module
- Level 4 — The Saber Grip Hacking Module and Spartan Module
- **Level 5 — The Reverse Grip Knife Stabbing Module**
- Level 6 — The Saber Grip Stabbing Module and Chain of the Knife Module
- Level 7 — The Pommel Strike and Closed Folder Strike Module
- Level 8 — In the Clutches Of
- Level 9 — Unarmed vs. the Knife
- Level 10 — The Knife/Counter-Knife Combat Scenario Module

CQCG SDMS (Single and Double-handed grip) Impact Weapon Course
- Level 1 — The SDMS Introduction and Quick Draw Combat Module
- Level 2 — The SMS Solo-Hand Grip Command and Mastery Module
- Level 3 — The DMS Double-Hand Grip Command and Mastery Module
- Level 4 — The SDMS CQC Block and Strike Combat Module
- **Level 5 — The SDMS Weapon Disarms and Retention Combat Module**
- Level 6 — The DMS Pull Grappling Series: Combat Module
- Level 7 — The DMS Push Grappling Series: Combat Module
- Level 8 — The DMS Turn Grappling Series: Combat Module
- Level 9 — The Unarmed Combatives vs. SDMS Attacks Module
- Level 10 — The SDMS Combat Scenarios Module

CQCG Gun/Counter-Gun Course
- Level 1 — The Quick Draw Combat Module
- Level 2 — The Walking Point/Search Module
- Level 3 — The Control, Restrain and Contain Arrest and Capture Module
- Level 4 — The Pistol Disarm and Retention Combat Module
- **Level 5 — The Long Gun Disarm and Retention Module**
- Level 6 — The Shoot / Move/Cover Module
- Level 7 — Emergency! The Tactical Medicine and Fight While Wounded Module
- Level 8 — Extreme Close Shooting Module
- Level 9 — You! Hostage! Module
- Level 10 — The Gun Combat Scenarios Module

CQCG

UNARMED COMBATIVES
The Level 5 Strike: The Punch/Counter-Punch Module

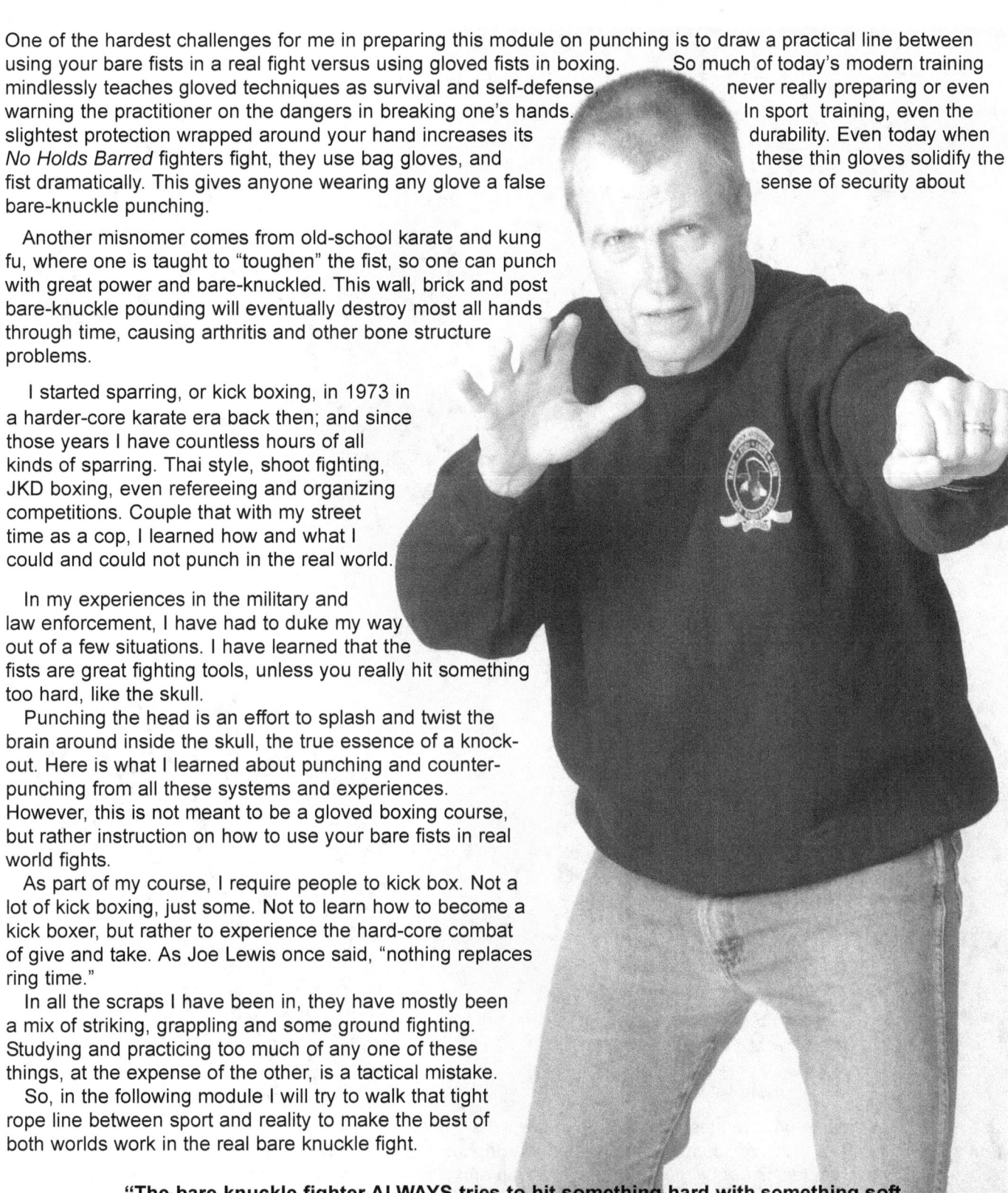

One of the hardest challenges for me in preparing this module on punching is to draw a practical line between using your bare fists in a real fight versus using gloved fists in boxing. So much of today's modern training mindlessly teaches gloved techniques as survival and self-defense, never really preparing or even warning the practitioner on the dangers in breaking one's hands. In sport training, even the slightest protection wrapped around your hand increases its durability. Even today when *No Holds Barred* fighters fight, they use bag gloves, and these thin gloves solidify the fist dramatically. This gives anyone wearing any glove a false sense of security about bare-knuckle punching.

Another misnomer comes from old-school karate and kung fu, where one is taught to "toughen" the fist, so one can punch with great power and bare-knuckled. This wall, brick and post bare-knuckle pounding will eventually destroy most all hands through time, causing arthritis and other bone structure problems.

I started sparring, or kick boxing, in 1973 in a harder-core karate era back then; and since those years I have countless hours of all kinds of sparring. Thai style, shoot fighting, JKD boxing, even refereeing and organizing competitions. Couple that with my street time as a cop, I learned how and what I could and could not punch in the real world.

In my experiences in the military and law enforcement, I have had to duke my way out of a few situations. I have learned that the fists are great fighting tools, unless you really hit something too hard, like the skull.

Punching the head is an effort to splash and twist the brain around inside the skull, the true essence of a knock-out. Here is what I learned about punching and counter-punching from all these systems and experiences. However, this is not meant to be a gloved boxing course, but rather instruction on how to use your bare fists in real world fights.

As part of my course, I require people to kick box. Not a lot of kick boxing, just some. Not to learn how to become a kick boxer, but rather to experience the hard-core combat of give and take. As Joe Lewis once said, "nothing replaces ring time."

In all the scraps I have been in, they have mostly been a mix of striking, grappling and some ground fighting. Studying and practicing too much of any one of these things, at the expense of the other, is a tactical mistake.

So, in the following module I will try to walk that tight rope line between sport and reality to make the best of both worlds work in the real bare knuckle fight.

"The bare knuckle fighter ALWAYS tries to hit something hard with something soft (i.e. palm-to-upper head) and something soft with something hard (i.e. fist-to-torso)."

Studies and Observations 1) The Four Main Fist Positions

The Vertical Fist.

The 45 Degree Angle Fist. The most natural.

The Horizontal Fist.

The Palm-up Fist.

These are the four main punch angles. Fists will be delivered at angles in between these as you, like a "heat-seeking" missile, try to strike a dodging head in and around the opponents limbs. A fighter should be free to deliver punches at all these angles. Every fist position has a reason and a maximized application that the others cannot offer. Taking one fist angle from a fighter's repertoire, as some systems mandate, is like taking a bullet from your gun. Go the field fully loaded.

Studies and Observations 2) The Rule of Thumb

When you teach new students, you discover they do not know where to put their thumbs. The thumb is always in danger of being hurt while punching. Some thumb positions are smarter than others.

The Flagpole Thumb. The sign that someone has no idea how to punch, or has worn a boxing glove too long.

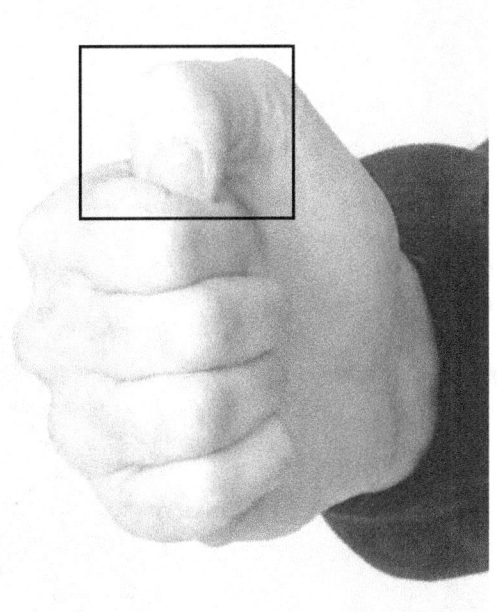

Some systems advocate the **Sun Fist Thumb**. This thumb is bent uncomfortably topside. Hit off target, and prepare for excruciating pain and thumb injury.

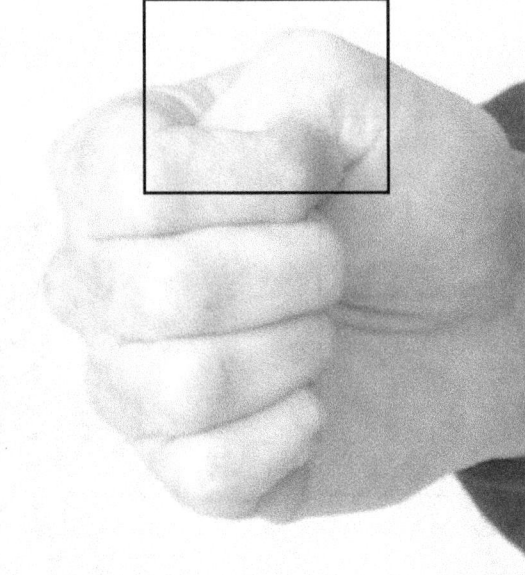

The Plugged Thumb. This prevents you from making a solid enough fist to punch safely.

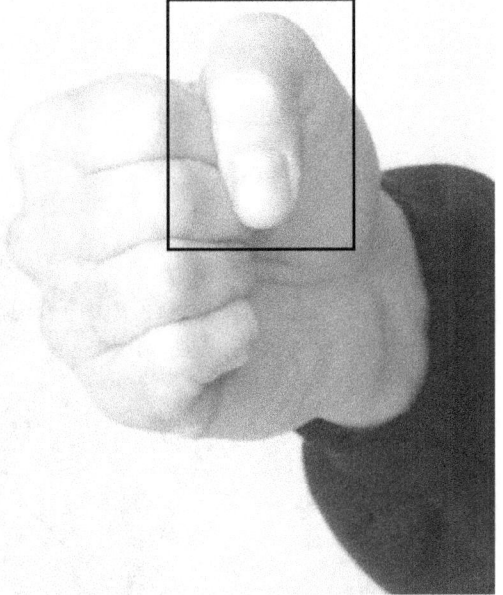

The winner by majority is **The Side Thumb**. It clears the fist and knuckles. It actually braces the fingers.

Upon impact, your fist must be as solid and tight as possible.

Studies and Observations 3) The Knuckle Controversy

Various systems mandate differing striking knuckles. The most popular bare knuckle choice is the top two knuckles. Positioning your fist, in conjunction with your wrist and forearm, is always a proper technique.

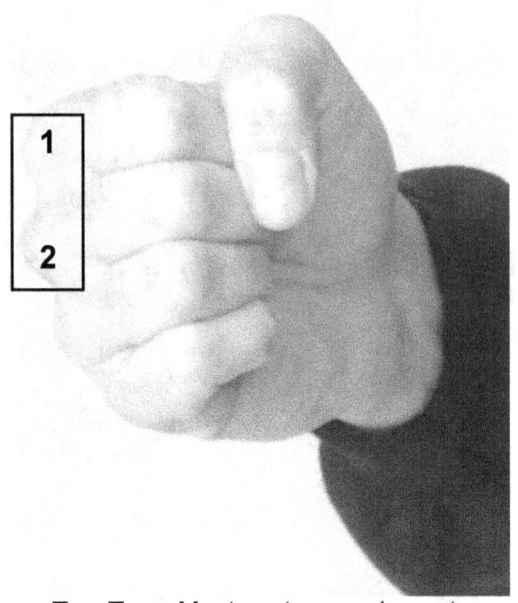

Top Two: *Most systems ask you to strike with the top-two knuckles. To produce the best, supported, straight alignment with the forearm, you learn to cant your fist forward.*

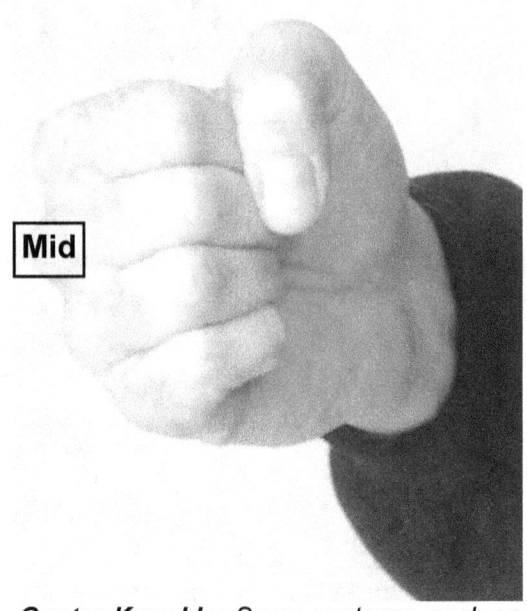

Center Knuckle: *Some systems, such as some Indonesian Silats mandate you strike with the middle knuckle. This knuckle is in the center and is easily aligned with the forearm.*

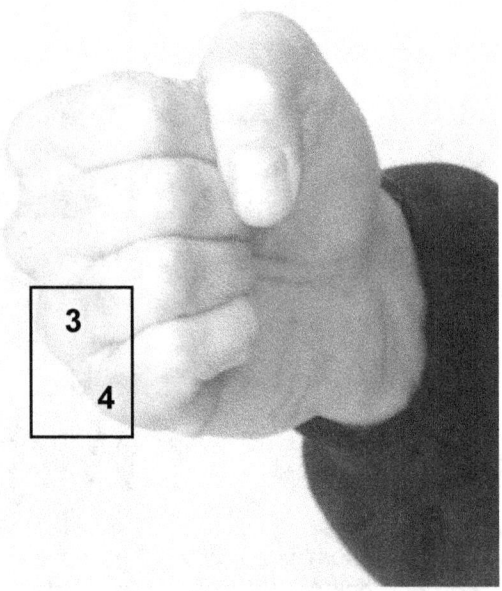

Lower Two: *Many systems, such as Chinese versions want students to punch with the lower two knuckles. This also must be supported by forearm alignment.*

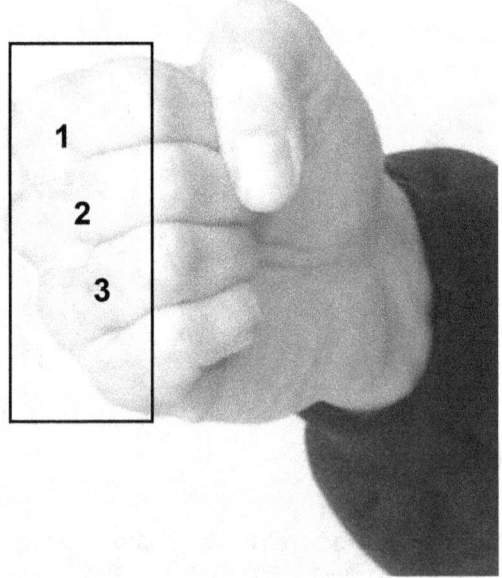

Dempsey Knuckles: *Boxers will say use the top three and the finger tops above them. But they are wearing gloves, and knuckles are lost inside gloves!*

Studies and Observations 4) The Other Hand: Sport versus Reality - A Comparison

Where and how you protect yourself with your free hand is important. Sport, gloved fighting has the tactic of placing the block up beside the head. This is easier on the hand and head when you have padded headgear, the thick padded glove. Plus, the opponent is wearing a thick padded glove. Do this same move in a gloveless street fight? The bare knuckle impact is quite substantial, and damaging to the back of the blocking hand.

There is a considerable difference in blocking punches with and without a glove on your blocking hand.

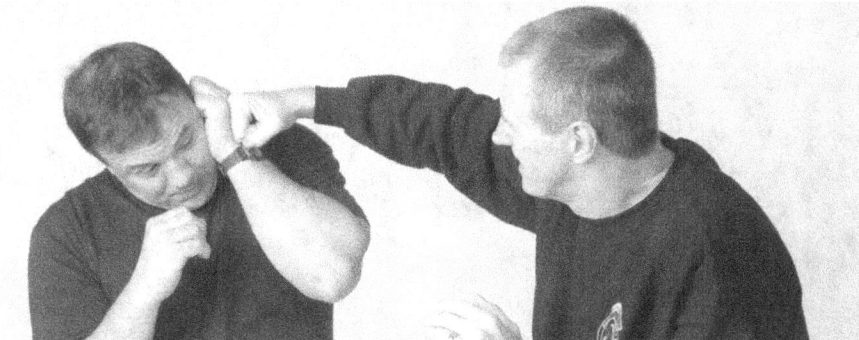

The Jab's Support Hand

The free hand is ready for anything.

We will cover the big five punches:

Punch 1) The Lead Jab
Punch 2) The Rear Cross
Punch 3) The Hook
Punch 4) The Uppercut
Punch 5) The Overhand

...and look instead at the position of the other hand, that is the one *not* punching. Instead of the blocking hand tucked against the head as many sport fighters with padded gloves do, the reality fighter keeps his hand up and ready in the window of combat.

The Cross Support Hand

The free hand is ready for anything.

Instead of the blocking hand tucked against the head as many sport fighters with padded gloves do, the reality fighter keeps his hand up and ready in the window of combat.

The Hook's Support Hand

Sport   **Reality**

The free hand is ready for anything.

Instead of the blocking hand tucked against the head as many sport fighters with padded gloves do, the reality fighter keeps his hand up and ready in the window of combat.

The Uppercut

Sport   **Reality**

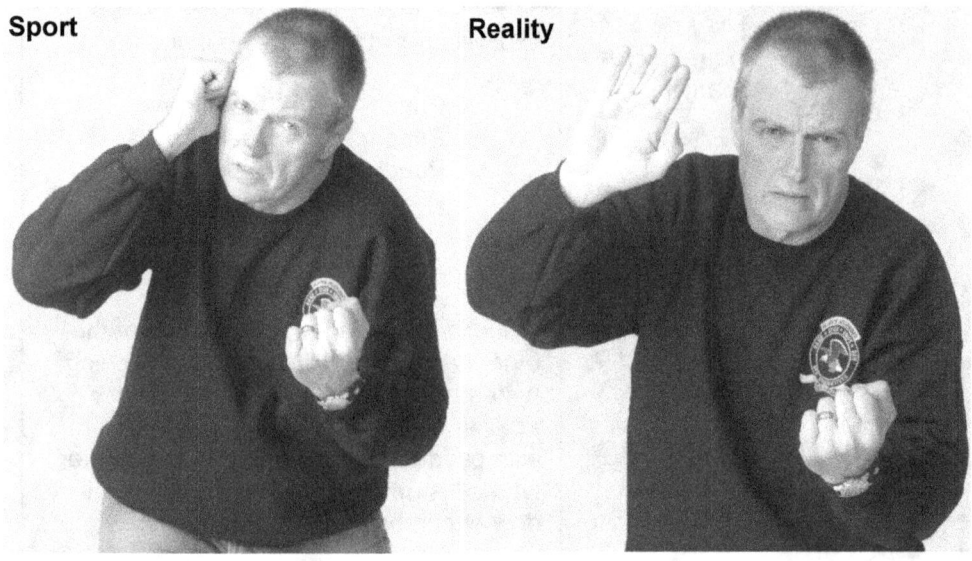

Instead of the blocking hand tucked against the head as many sport fighters with padded gloves do, the reality fighter keeps his hand up and ready in the window of combat.

The Overhand

Sport   **Reality**

Instead of the blocking hand tucked against the head as many sport fighters with padded gloves do, the reality fighter keeps his hand up and ready in the window of combat.

Studies and Observations 5) A Crash Course on the Basic Punches

As a survival fighter, I am reluctant to call a right or a left punch the typical sport names like jabs or crosses, yet these terms have become so universal, they help in quick communication. Without making this a chapter on sport boxing, here is a working knowledge of the basic punches. It is important that I include some sport punching information so as to explain the differences between sport and reality. So many instructors blindly confuse the two out of sheer ignorance. Remember that all strikes and targets deemed illegal in sport boxing must be used in street fighting. We will start with what is typically called the jab punch.

The Jab Punch:
Fighting stance/position
The body is naturally bladed in a fighting position which includes slightly bent knees, with shoulders up and chin down. The jab punch is fired from the lead shoulder or, in layman's terms, the closest shoulder to the enemy.

A street fighter tries to develop both sides in this mixed weapon world of striking and grappling atop the jagged landscapes of real life. The general advice is to develop body synergy with the strike. Try and bring your shoulders and torso in with the punch. Sometimes you might even take a step. All this is developed by striking heavy bags and focus mitts.

The so-called jab punch may also be a lead leg/shoulder punch to the torso.

The Five Major Punches

1) The Jab

2) The Cross

3) The Hook

4) The Uppercut
   - tight uppercut
   - extended uppercut

5) The Overhand
   - forearm/fist strike
   - hooking style strike

Straight or cork-screw delivery
Many boxers think that the Jab is the most important strike in a Boxers arsenal. But this may or may not be true in a street fight. When throwing a jab, sport boxers make sure they twist your arm in a cork-screw like motion just before impact. They feel the cork-screw is what will give you the twisting snap needed for a good punch. The punch lands in horizontal or a thumb-down more than with a horizontal impact. The sport boxing cork-screw motion also facilitates a skin tear as the gloves may rip open the flesh. However many people are taught to jab without the cork-screw and with a vertical fist only, completely ignoring the cork-screw. You should practice and understand both as you will need the ability to chase your dodging target.

The Hammer Jab - *taught to me by JKD expert Tim Tackett, has a two-fold mission. Your forearm hits his guarding arm to clear a path for the fist to hit the chin or throat. This may make the striking arm "chop" down onto the target.*

Extended or snapping
The jab may blast in with deep penetration and full body momentum. This means the arm stays out and extended longer and may be susceptible to a counter-attack. The jab may snap back. After penetration, it may yank back as fast or faster than it went out. Some describe this type of jab as akin to snapping a towel.

The street fighter jab has several duties

1) Power Jab - strong enough to hurt the whole body or as a knock-out.

2) Probing Jab - to check distance and reaction for possible set-ups and fakes.

3) Shielding Jab - to keep an enemy back and away.

4) Hammer Jab - that breaks down his reflexive guard while striking.

The Cross Punch:
Fighting stance/position
The body is naturally bladed to one side in a fighting position, which should include slightly bent knees. Shoulders are up and chin down. The cross punch is fired from the rear shoulder or, in layman's terms, the farthest shoulder from the enemy. A trainee tries to develop both sides. You will always have a strong-side but may not be able to fight from this favored position. The general advice is to try and feel your shoulder and hip to some extent being thrust into the punch. Sometimes fighters take a step with the strike. All this is developed by striking heavy bags and focus mitts.

The so-called cross punch may also be a rear leg/shoulder punch to torso.

Randy's right punch is crossing his chest, the boundary of his left shoulder and the distance to his target, my chin.

Note his right leg is back. All these are defined elements of a classic cross punch.

A cross punch can be high, or low as targets open up.

The Delivery
The cross punch is similar to the jab punch. In sports, the cross is usually set up by the jab, but in reality fighting it may not be.

The starting position and the motion is the same except the trailing hand, shoulder and hip are used instead of the leading hand, shoulder and hip.

Many instructors suggest pivoting the rear foot. Others mandate dropping the heel on the floor for full support. You practice both and decide for yourself.

Above, a rear, right cross punch may shoot right between both of Randy's arms. If I happened to be in a left lead, the punch would then be a front leg, left punch. Learn to punch and fight from both left and right leads.

The Hook Punch:

Fighting stance/position

The body is naturally bladed in a fighting position which includes slightly bent knees, shoulders are up and chin down. The hook punch is fired from either side of the body in combat. The general advice is to try and feel your shoulder and hip to some extent being thrust into the punch. All this is developed by striking heavy bags and focus mitts. The hook punch may hit the jaw, neck and torso. Even a steady pounding on the opponent's arms may take a toll.

Delivery

A hook is a punch executed in a tight arc from the outside of the body to the inside in a hooking motion. A hook is a very powerful and slightly deceptive punch if executed correctly. It can be thrown by either arm and almost any position, it is usually practiced within a combination of punches. It is usually used to hit the torso, neck and head. It involves a slight rotation on the ball of at least one foot.

Loose Hooks and Roundhouse Punches

The classic, old-fashioned "haymaker punch" is a hooking strike that is useful for punching around blocks or guarding arms that would otherwise prevent a straight punch from hitting its target. It is also useful against an opponent who turns and ducks. A loose hook barely looks like a hook punch. A roundhouse haymaker punch looks like an exaggerated hook punch. Haymakers are said to be the most common street punch.

A Tight Hook Punch

A tight hook is perfect for close quarters and ground fighting, for times when there is not a lot of space for arm movement.

The "Loose Hook." A long range, straight punch may hunt down a dodging and ducking head and in this pursuit often turns into a hooking punch. Note the fist position.

"The Medium Range" hook punch.

The Hook Punch Summary

Long Range Loose Hook.
  - Very slight hooking to chase moving targets.

Medium Range.
  - The enemy is in closer.

Tight Hook for very Close Quarters.
  - The enemy is very close.

Hooks are thrown standing and on the ground.

Hooks are right or left handed, lead or rear arms.

The higher the hook, the more the fist should turn from vertical to a horizontal position.

Hooks should be delivered with body synergy.

The Tight Hook:
Here is an example of a tight hook punch. Two fighters crash, then clinch. Your punch goes up, elbow rises, and the punch goes in.

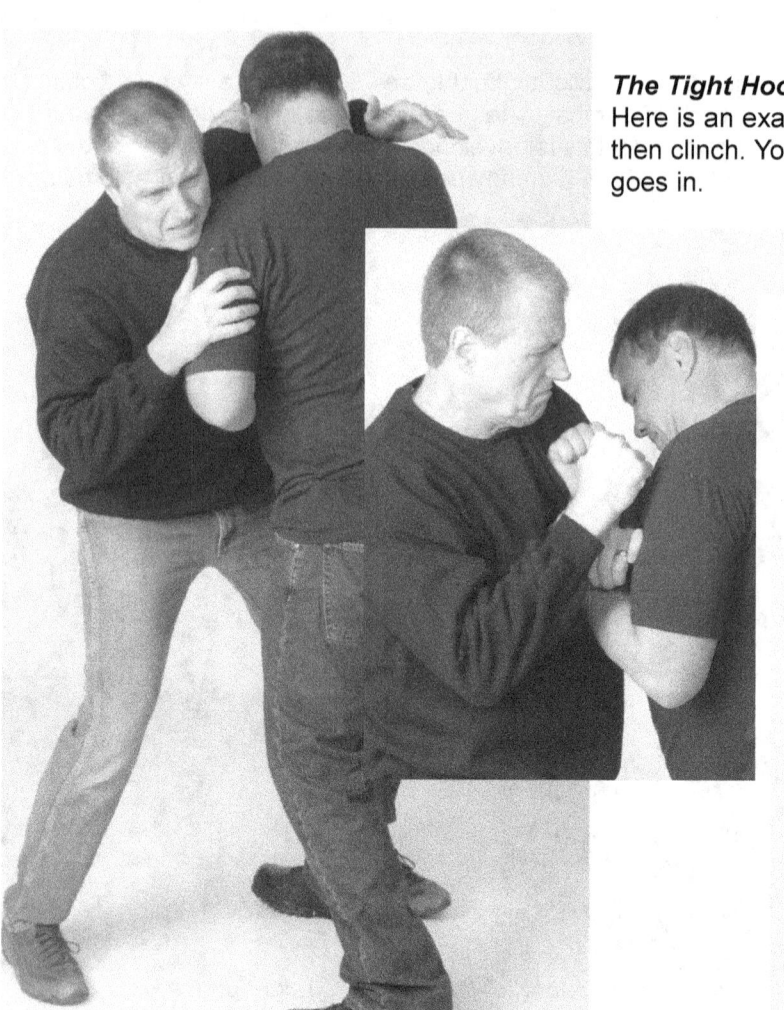
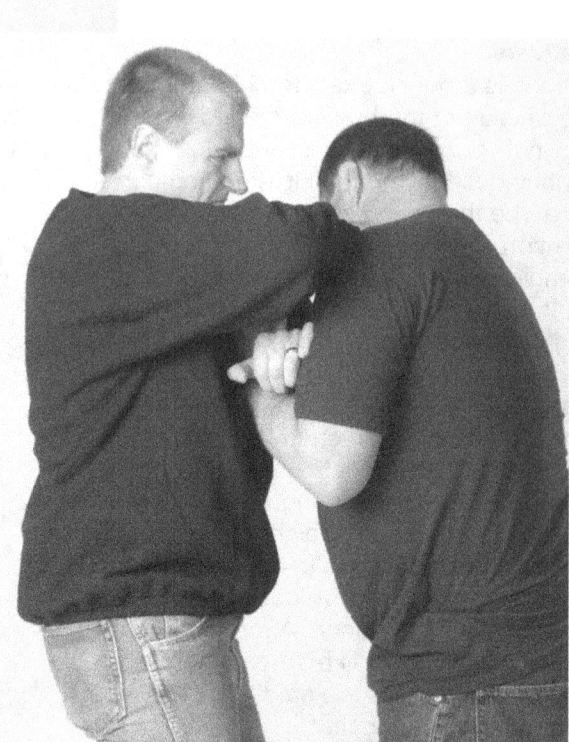

Here's the tight hook again "in the air." The fist goes into position. The elbow rises and strikes. Get solid penetration. It should retract immediately. This is often a good punch in very close-quarters.

High or low, tight or not, do not over-strike your hook. It sets you up for counters. Hit the target and retract back to a cover.

The Uppercut Punch:
Fighting stance/position
The body is naturally bladed in a fighting position, which includes slightly bent knees. Shoulders are up and chin down. The uppercut punch is fired from either side of the body in combat. The general advice is to try and feel your shoulder and hip to some extent being thrust into the punch.

The Delivery
The uppercut begins with the punching hand near the waist and the elbow pointed backward. You bend the knees and explode upward using your legs to generate power. As you punch, the fist moves upward as the arm rotates at the shoulder with the elbow remaining bent and brushing along the side as the fist moves upward. Keep the elbow under the fist with the wrist held straight. The fist may be used with or without fist rotation. The uppercut must be coordinated with the hips and legs to generate maximum power.

Uppercuts are delivered to targets that are tilted forward, such as an opponent leaning forward after a body shot, exposing his torso or his chin. However, an opponent who is standing upright may furnish a target if the chin is protruding forward.

The Uppercut Punch Summary

Tight Hook for very Close Quarters.
- The enemy is very close.

Extended Range.
- The enemy is not so close.

Are thrown standing and on the ground.

Are right or left handed, lead or rear arms.

Should be delivered with body synergy.

Do not overshoot the target. Hit it and retract.

The Tight Uppercut usually to chin. *The Extended Uppercut to chin, throat or torso.*

Set-up and then an extended reach uppercut. A sample of the "Shovel Hook" uppercut, or what is sometimes called the "Mexican Uppercut." These are shots to the torso.

When in serious, lethal-force encounters, these extended uppercuts to the throat are very successful.

The Overhand Punch:
Fighting stance/position
The body is naturally bladed in a fighting position, which includes slightly bent knees. Shoulders are up and chin down. The overhand punch is fired from either side of the body in combat. The general advice is to try and feel your shoulder and hip to some extent being thrust into the punch.

The Delivery
The arm raises and descends upon the opponent. You either strike his forearm clear, or strike around his forearm in what are often called *Descending Overhands* and *Hooking Overhands*.

The Descending Overhand
This high shot drops down on the opponent's lead protecting arm. The dropping forearm hits the defending forearm as you try with your fist to reach out and strike his jaw and neck.

The Hooking Overhand
With this punch descend downward; and with the fist find an opening to work around and strike the jaw or neck.

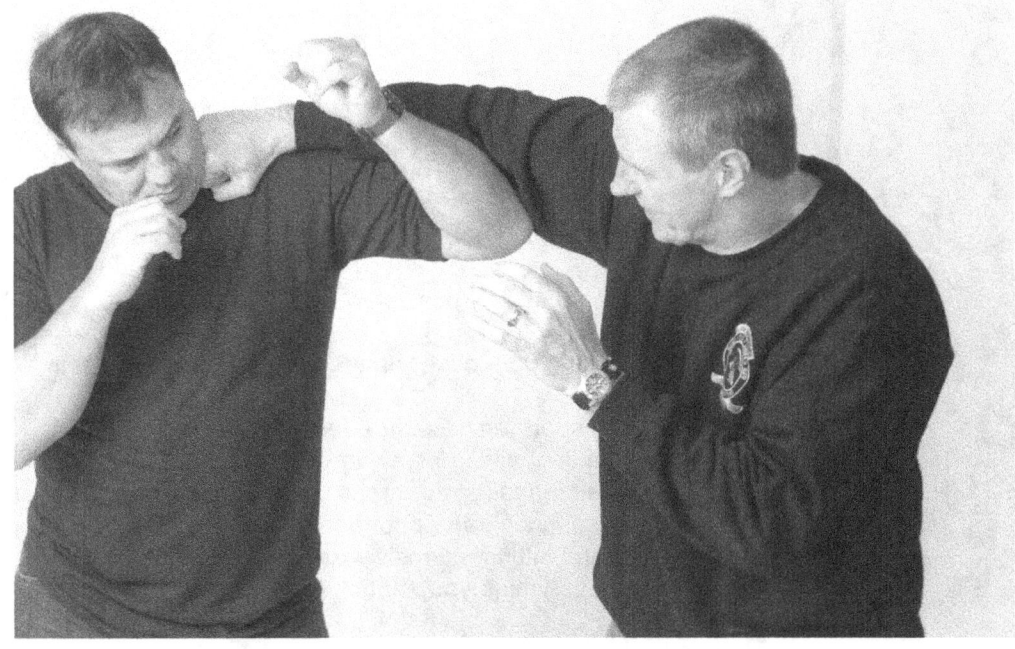

The "Skinny Fist":
This has been called numerous things in many systems, often as some kind of animal paw. It's a viable fist formation and allows for a strike to the throat and neck. The strike can hit from a thrust or a hook delivery. Be advised that a windpipe attack may well be lethal.

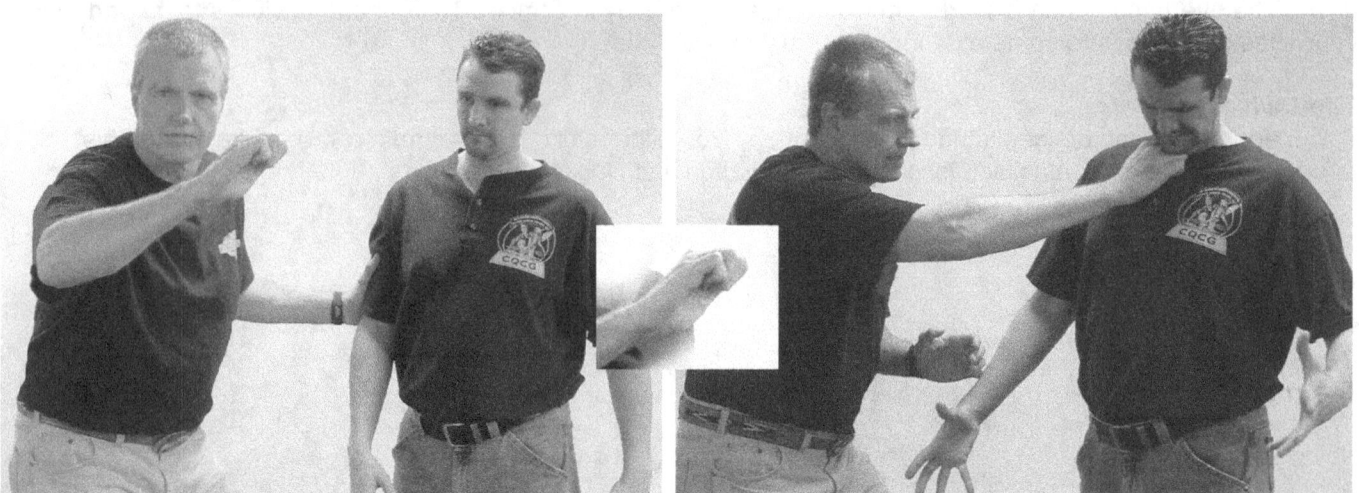

Survival Punching Summary

Punching with your closed fist is but one of many striking tools.

Sport boxing and survival punching do have some of the same strategies.

Use total body synergy to support each strike.

Practice hitting a heavy bag.

Practice hitting focus mitts.

Aim past the surface of your target to ensure penetration.

Aim low on the face to avoid hand injury when hitting the skull.

Classical hand toughening can lead to long-term injury and arthritis.

Support whatever knuckles you use to hit with proper arm alignment.

At the point of impact, tighten your fist to the extreme.

Use straight punches and hooking punches from your right and left hands.

Protect yourself as much as possible.

Beware of positioning your cover hands too close to your head.

Dirty Boxing is a sport term used to describe anything foul, such as an infringement of boxing rules, including: hitting below the belt; hitting with any part of the body other than the knuckles; leaning against the ropes; head-butting; hitting the back of the opponent's neck, head or torso; hitting an opponent who is down; throwing a punch while in a clinch; holding; holding and hitting; offensive language; assaulting or action; tripping; kicking. In short, just about everything you should do in a street fight.

The Double Punch:
Here, for your review is the lost or forgotten tactic of the double punch. This is a tactic from various classical martial arts that has a powerful and devastating advantage, if used at the right moment. If one arm is not needed for a second for protection, the double punch can be applied. I include it here as it is a dynamic tool and I hope to revitalize it.

Solo Command and Mastery of the Double Punch

Top hand at 12 o'clock.

Right hand at 3 o'clock.

Bottom hand at 6 o'clock.

Left hand at 9 o'clock.

One Double Punch Scenario Sample

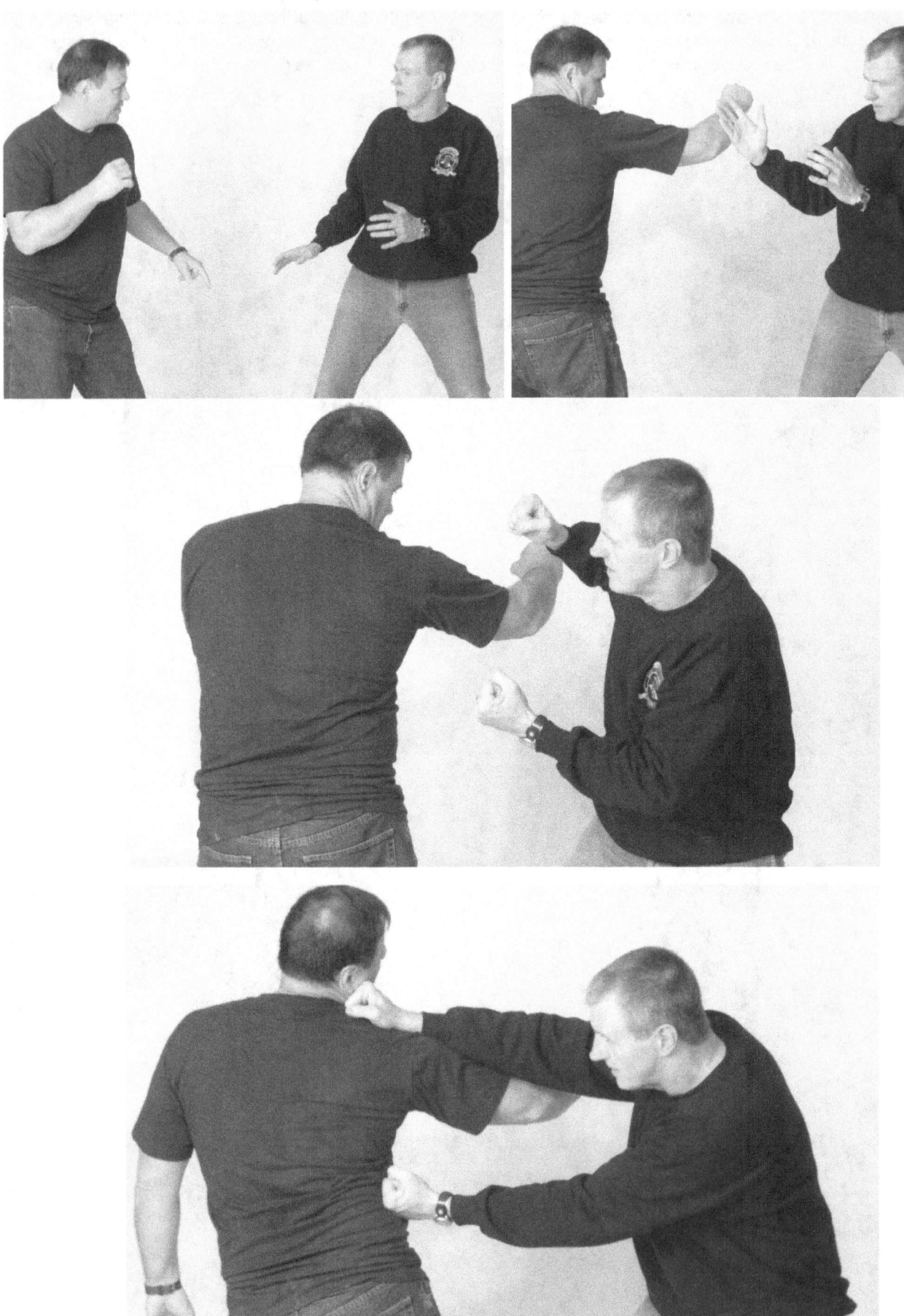

Page 26 - W. Hock Hochheim's Training Mission Five

Your Solo Command and Mastery Punching Workout

Do in the air. Treat this like shadow-boxing. Imagine hitting an enemy. Hit a heavy bag.

Basic Training Set:
10 Jabs right hand: a) head shots, b) body shots
10 Jabs left hand: a) head shots, b) body shots

10 Crosses right hand: a) head shots, b) body shots
10 Crosses left hand: a) head shots, b) body shots

10 Right jabs and left crosses
10 Left jabs and right crosses

10 Hooks right hand - high
10 Hooks left hand - high
10 Hooks right hand - low
10 Hooks left hand - low
10 Hooks right - tight
10 Hooks left - tight

10 Right jabs, left crosses and right hooks
10 Left jabs, right crosses and left hooks

10 Uppercuts right extended
10 Uppercuts left extended
10 Uppercuts right tight
10 Uppercuts left tight

10 Right jabs, left crosses, right hooks, left uppercuts
10 Left jabs, right crosses, left hooks, right uppercuts

10 Overhand hooks, curve right
10 Overhand hooks, curve left
10 Overhand descenders right
10 Overhand descenders left

10 Right jabs, left crosses, right hooks, left uppercuts, right overhands
10 Left jabs, right crosses, left hooks, right uppercuts, left overhands

Advanced Training Set:
Varied combinations of above. We cannot list all the possibilities here. You must make your own list. Samples are:

10 Right uppercuts, left hooks
10 Left uppercuts, right hooks
10 Double hooks
10 Double uppercuts
10 Right overhands, left uppercuts
10 Left overhands, right uppercuts
10 Sets, right and left punches, then uppercut
10 Sets, right and left punches, then hook
10...you get the idea. Continue to build your personal list.

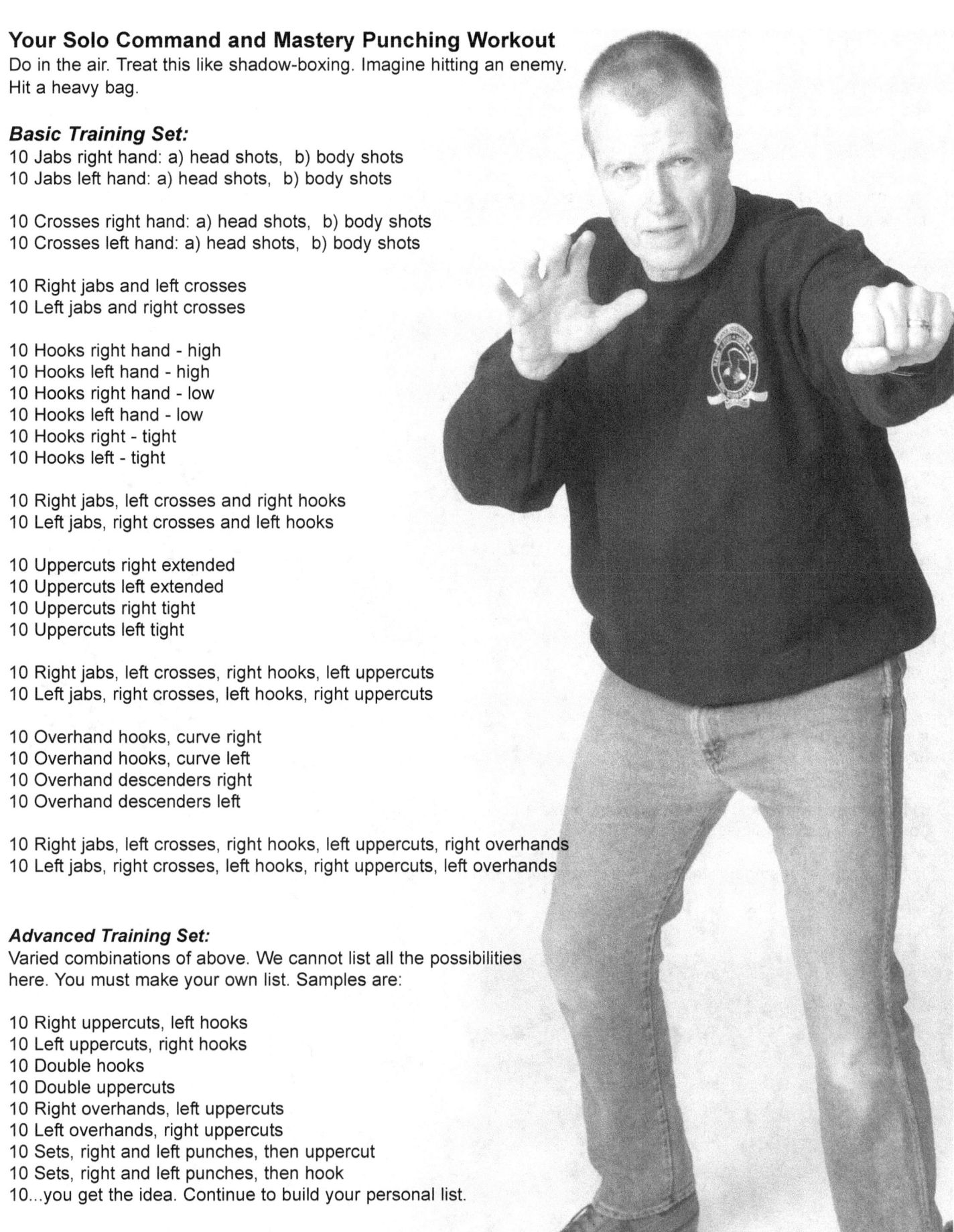

Partner Drills

Training the Punches

I do not want to create a sport boxer, so I will not display all the common boxing drills here, just the ones that I feel are important for the street or battlefield fight. This means doing some focus mitt drills, some contact drills, and some boxing. Mix these boxing drills with the kicking drills, and you have started kick boxing. The fighter becomes diminished. Mix the kick boxing with the takedowns and throws. Mix this with the street ground fighting, and you have an overall basis for reality fighting. As you proceed, please remember this and remember where the punching fits in the big picture. Over-emphasizing any one group is not tactically sound.

Focus Mitt Drills

Focus mitts are more for speed development and less for power. You develop power on training objects such as the heavy bag. Proper mitt feeding is both an art and a science. Some quick rules.

1) Keep the "face" of the mitt hidden as much as possible.

2) Keep the mitt near your head and body as safely as possible to remind the fighter where the target really is.

3) Leave the mitt up? This means he hits the exposed mitt multiple times. Do not be sloppy about this.

4) Do not position the mitt beyond his realistic contact point.

5) The faster you remove the exposed surface, the faster he learns to strike to hit it. This develops more speed.

6) Be creative with mitts and your workout. Move and simulate reality the best possible. Punch at him with a horizontal mitt. Sometimes hit him with the face of the mitt. Kick at him. Clinch him.

7) Do not create a student who is a "*focus mitt hitting expert.*" You are supposed to be creating a "*body hitting*" expert.

Basic Feed and Hit Focus Mitt Drills

Basic Training Skill Sets:
10 Jabs right hand
   a) head shots, b) body shots

10 Jabs left hand
   a) head shots, b) body shots

10 Crosses right hand
   a) head shots, b) body shots

10 Crosses left hand
   a) head shots, b) body shots

10 Right jabs and left crosses
10 Left jabs and right crosses

10 Hooks right hand - high
10 Hooks left hand - high
10 Hooks right hand - low
10 Hooks left hand - low
10 Hooks right - tight
10 Hooks left - tight

10 Right jabs, left crosses and right hooks
10 Left jab, right crosses and left hooks

10 Uppercuts right extended
10 Uppercuts left extended
10 Uppercuts right tight
10 Uppercuts left tight

10 Right jabs, left crosses, right hooks, left uppercuts
10 Left jabs, right crosses, left hooks, right uppercuts

10 Overhand hooks, curve right
10 Overhand hooks, curve left
10 Overhand descenders right
10 Overhand descenders left

10 Right jabs, left crosses, right hooks, left uppercuts, right overhands
10 Left jab, right crosses, left hooks, right uppercuts, left overhands

Advanced Training Set:
Varied combinations of the above. Numerically, we can list all the possibilities here. You must make your own list. Samples are:
10 Right uppercuts, left hooks
10 Left uppercuts, right hooks
10 Double hooks
10 Double uppercuts
10 Right overhands, left uppercuts
10 Left overhands, right uppercuts
10 Sets, right and left punches on one mitt, uppercut on the other
10 Sets, right and left punches one mitt, hook punch on the other
10 Sets, trainer holds a shield sideways, throws up a guard arm as an obstacle.
10....You get the idea. Build your personal list.

Straight punches (jabs or crosses).

Straight punches (jabs or crosses).

Hook punches.

Defeat-a-guard arm punches.

Slap Mitt Drills

These mitt slap drills are my favorite focus mitt drills. They were first inspired by Larry Hartsell in the 1980s, and I have worked and organized them into various drills. For starters, the student needs to know two basic crunch blocks (see earlier *Training Mission* Books for these details) and the jab, cross, hook and uppercut punches:

The Shoulder Crunch   The Torso Crunch   The Jab

The Cross   The Hook   The Uppercut

These give-and-take slap mitt drills are based on hitting the trainee with the mitt first and then re-positioning the mitt so the trainee can counter-punch and return fire. This toughens the trainee and, at the same time, is somewhat safe training.

The trainee takes the blow with a slight sidestep and shift of the body. Against a downward attack, he raises his shoulder, drops his chin and takes the mitt with a shoulder-crunching block. His hands may go anywhere, but the high-arm, low-arm coverage, or zone coverage, is probably the best. Against a lower attack, the trainee braces his arm against his torso, arms in the best position for the second.

The trainer must remember to re-position the mitt in a realistic manner, else the trainee will aiming at a false point of contact.

The Hook Punch Slap Mitt Drill

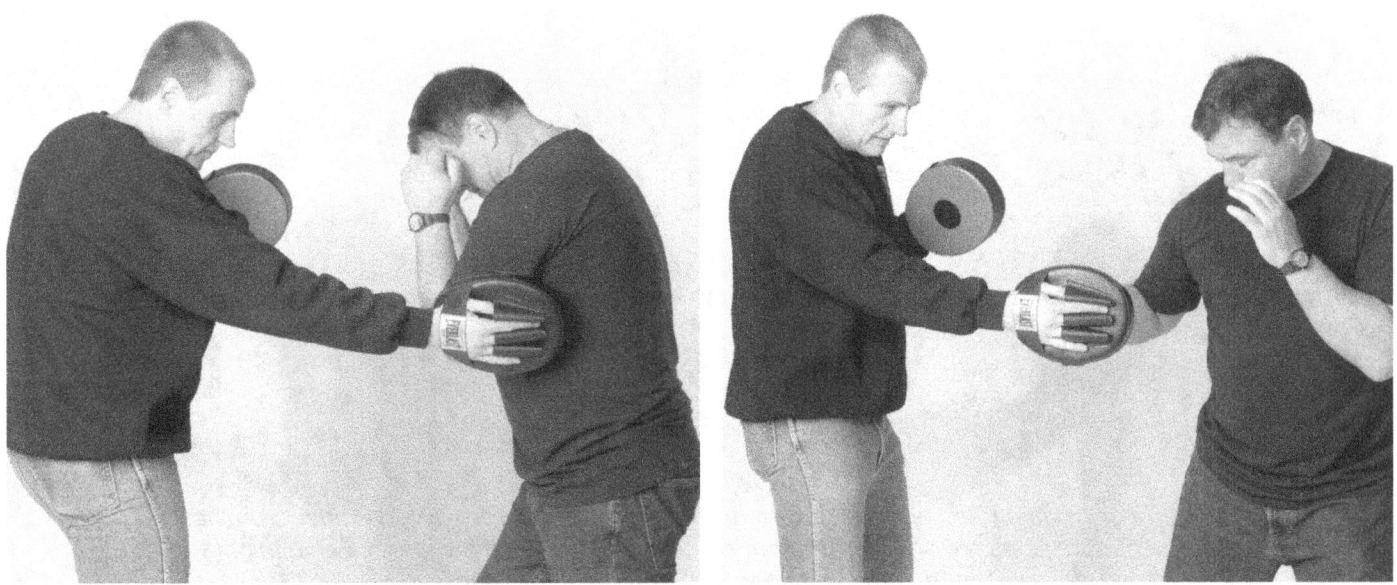

The trainer hauls off and hits the trainee hard on his left side. He torso crunches. The trainer positions the same mitt about the center-line of his body. The trainee strikes the mitt with his right hook.

The trainer hauls off and hits the trainee hard on his right side. He torso crunches. The trainer positions the same mitt about center-line of his body. The trainee strikes the mitt with his left hook.

The trainer hauls off and hits the trainee hard on his left side. The trainee torso crunches. The trainer positions the same mitt about center-line of his body. The trainee strikes the mitt with his right hook. The trainer hauls off and hits the trainee hard on his right side. The torso crunches. The trainer positions the same mitt about center-line of his body. The trainee strikes the mitt with his left hook.

"Give him some flak"

Work this in sets of 10. One complete set is four beats as shown above. Begin to add some distracting moves I like to call "flak," like poking him with the top edge of the mitt. Trainees quickly understand the flak inside the basic movements of the drill.

Note: I usually start teaching the slap mitt hook drill first because it is easy for a new student to see, learn and understand the slap format. Then I proceed to the other punches.

The Uppercut Slap Mitt Drill

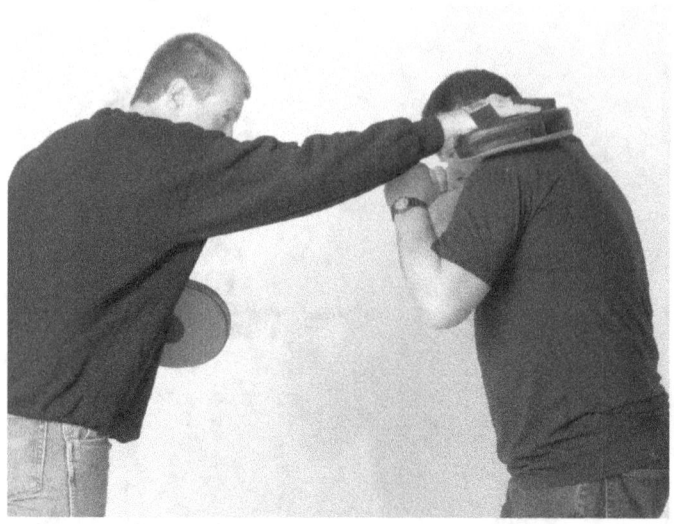

The trainer hauls off and hits the trainee hard on his left shoulder. He shoulder crunches. The trainer positions the same mitt about center-line of his body where his chin might be. The trainee strikes the mitt with his right uppercut.

The trainer hauls off and hits the trainee hard on his left shoulder. He shoulder crunches. The trainer positions the same mitt about center-line of his body where his chin might be. The trainee strikes the mitt with his right uppercut.

The trainer hauls off and hits the trainee hard on his left shoulder. He shoulder crunches. The trainer positions the same mitt about center-line of his body where his chin might be. The trainee strikes the mitt with his right uppercut. The trainer hauls off and hits the trainee hard on his right shoulder. He shoulder crunches. The trainer positions the same mitt about center-line of his body where his chin might be. The trainee strikes the mitt with his left uppercut.

"Give him some flak"

Work this in sets of 10. One complete set is four beats as shown above. Begin to add some distracting moves I like to call "flak," like poking him with the top edge of the mitt. Trainees quickly understand the flak inside the basic movements of the drill.

The Straight (Jab/Cross) Slap Mitt Drill

The trainer hauls off and hits the trainee with a horizontal mitt poke hard into his right shoulder. A trainer could also slap the shoulder with a horizontal mitt. The trainee gives at the shoulder. The trainer positions the same mitt about center-line of his body where his chin might be. The trainee strikes the mitt with his left straight punch.

The trainer hauls off and hits the trainee with a horizontal mitt poke hard into his left shoulder, or even the top of his head. A trainer could also slap the shoulder with a horizontal mitt. The trainee gives at the shoulder. The trainer positions the same mitt about center-line of his body where his chin might be. The trainee strikes the mitt with his right straight punch.

This drill blends the jab and cross into just a straight punch. You may position the trainee into differing leads if you wish to enforce a jab or a cross punch response.

The trainer hauls off and hits the trainee with a horizontal mitt poke hard into his right shoulder. A trainer could also slap the shoulder with a horizontal mitt. The trainee gives at the shoulder. The trainer positions the same mitt about center-line of his body where his chin might be. The trainee strikes the mitt with his left straight punch. The trainer hauls off and hits the trainee with a horizontal mitt poke hard into his left shoulder. A trainer could also slap the shoulder with a horizontal mitt. The trainee gives at the shoulder. The trainer positions the same mitt about center-line of his body where his chin might be. The trainee strikes the mitt with his right straight punch.

"Give him some flack"

Work this in sets of 10. One complete set is four beats as shown above. Begin to add some distracting moves I like to call "flack," like poking him with the top edge of the mitt. Trainees quickly understand the flak inside the basic movements of the drill.

Slap Mitt Ground Fight Applications

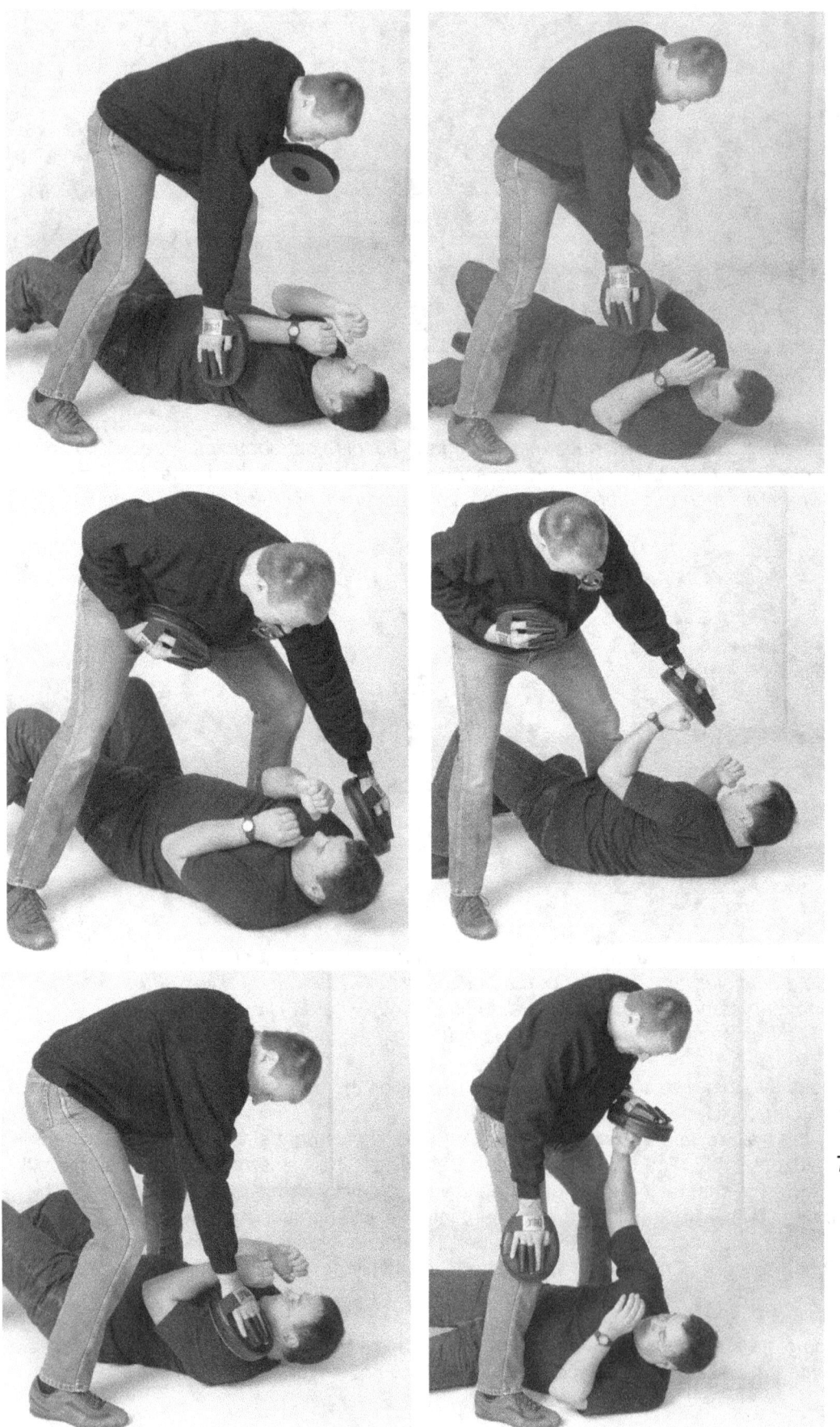

The Hook Punch version

The Uppercut version

The Straight Punch version

The Startle Slap Mitt Drill

The trainer stands before the trainee. The trainee's arms are down. The trainer suddenly screams, curses and yells then, with great force, slaps one or both mitts on the trainee's chest. This is a way to prep trainees for a madman attack and feel some body impact along the way. The trainee hits back in any series of strikes.

Slap Mitt Drills
Trainee stands:
 - with his arms down.
 - with his arms up and ready.
 - or on the ground.
   1) Straights
   2) Hooks
   3) Uppercuts
   4) Startles - all of the above.

Note - Thanks to Larry Hartsell for passing these on. As he requested, I pass them on...

Other Support Drills

The Peek-a-Boo Drill: Develop the Snap
Not a very macho name for an old school punching drill done by all the greats, to enhance speed.

First, I check the range of Randy's punch so as to ensure that he will not nail my nose in the drill! I stand before Randy and I assume the "prayer" position.

When I open my hands, he fires his hand in the opening. I will try to clap down on the punch to try and catch his fist. This teaches him to snap and punch even faster.

The So-Called "Perfect Punch" Drill

They say that looseness facilitates speed. The punch should start out with a loose hand/forearm/upper arm. As the punch explodes in, the fist tightens. The fist is tight on impact, then penetrates at least two inches and then snaps back, loosely. This is an important punching theory to be familiar with. You will find your own balance of looseness and tightness while training punches. Remember, you cannot always predict the exact distance and position of your enemy.

The Broken Hand
Most fists are chipped, fractured and/or broken when they hit the skull. In fact, a fast-ditch block of old bare-knuckle boxers was to lower the head and let the other paluka punch the skull. This often incapacitates the hand, or even the fighter himself. This head lowering movement is also a reflexive move by the untrained as well as the trained. Your bare fist is quite likely to be hurt. This harsh reality holds true whether you are doing the multiple strike, battle punch of Wing Chun or Jeet Kune Do, the power karate punch, or just brawling swinging fists. It is a crime of ignorance when these systems fail to warn their students of this typical problem.

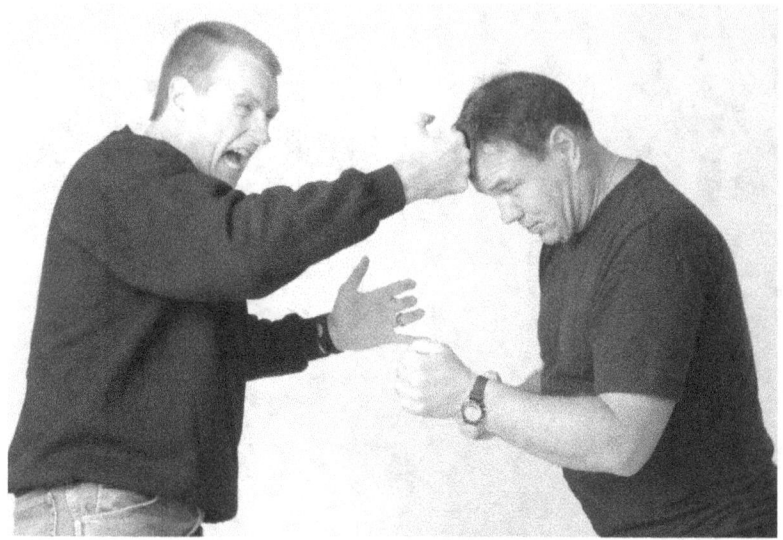

Bigger Bones Win?
Medical experts report that a small bone will break and/or dislocate before a big bone, and I think that is true. You can look at person's hands and predict whether they should primarily be punchers or not. If the hands look "fine," if the knuckles and joints are boney? As compared to those hulky, meaty hands we see people have. A good instructor should oversee this when a student starts to build their own 6 to 12 survival tricks, But if the student wants to start teaching? He has to work on and know the meaty hand material too!

Many of my karate veteran friends have broken their hands aiming at the nose. Many concept fighters have busted their hands doing the battle punching taught and so easily done on focus mitts, so easily broken by a ducking, dodging head. Modern approaches suggest aiming for the jaw.

If you must punch at the head area, aim at the neck or jaw. The neck is soft and a punch to the carotid is a great target. Windpipe punches may be deadly, so you had better be in a lethal force necessary situation.

A punch to the jaw line is a sound idea because the jaw moves and head twists. The skull is a culprit that breaks your hand. Next is the forehead which has caused me to need hand surgery. Next, hitting the ocular cavity around the eyes can be sharp and dangerous to strike. Some severe cheekbones have caused damage. All this, when supported on a stout neck, leads to a damaged hand.

The head twists on the neck, the jaw gives and both may soften the bone impact on your fist.

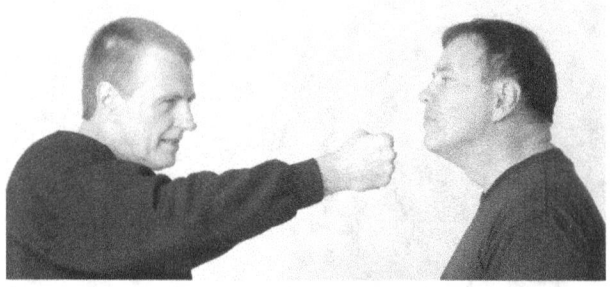

Front punch? Aim low for the chin or neck. If the head drops? You get the nose.

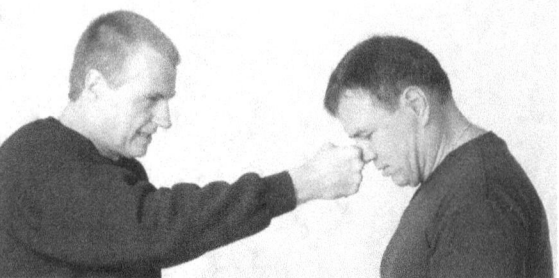

The British Commando Punching Drill
This drill comes from the World War II days, when the British troops realized the dangers of punching at the nose and head. The enemy reflexively lowers the head, exposing the bare fist to the hard forehead and skull. The drill emphasized three safer targets, the jaw (which is moveable) the torso and the groin. And of course, the palm heel strike was a preferred strike to save the closed fist hand from injury.

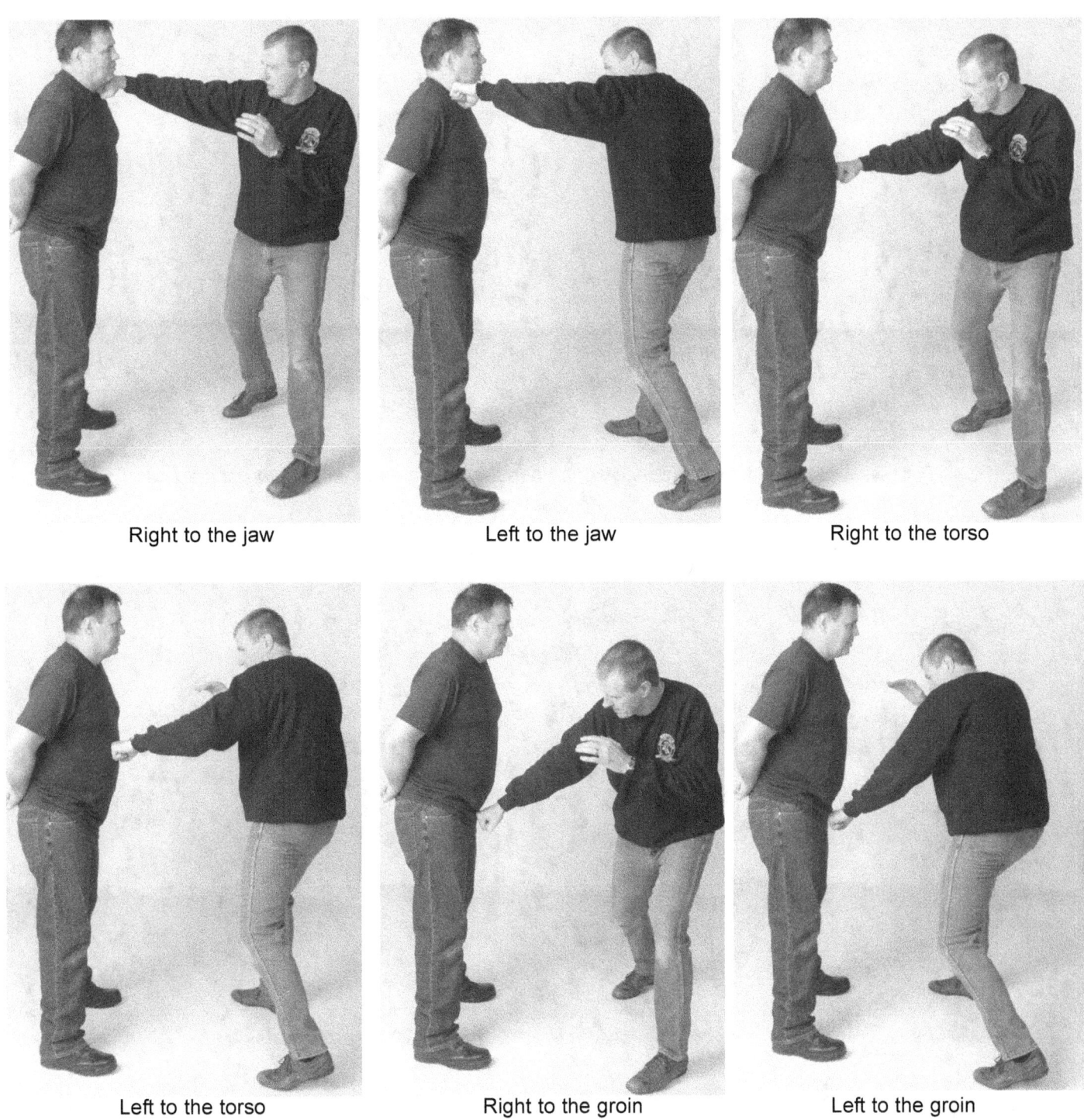

Right to the jaw Left to the jaw Right to the torso

Left to the torso Right to the groin Left to the groin

Punching is such a natural move, militaries of the world have tried to create some muscle memory with this drill and guide the strikes to the least harmful targets to the hands. Hit something hard with something soft. Hit something soft with something hard.

The Wall Drill

This is a drill I think I have invented, or at least I have never seen it before anywhere around the world. It sets training partners a safe distance apart and develops blocking skills. Due to the set-up, the trainee experiences stress because he sees these blows flying at him at realistic speeds, and they come VERY close to hitting him.

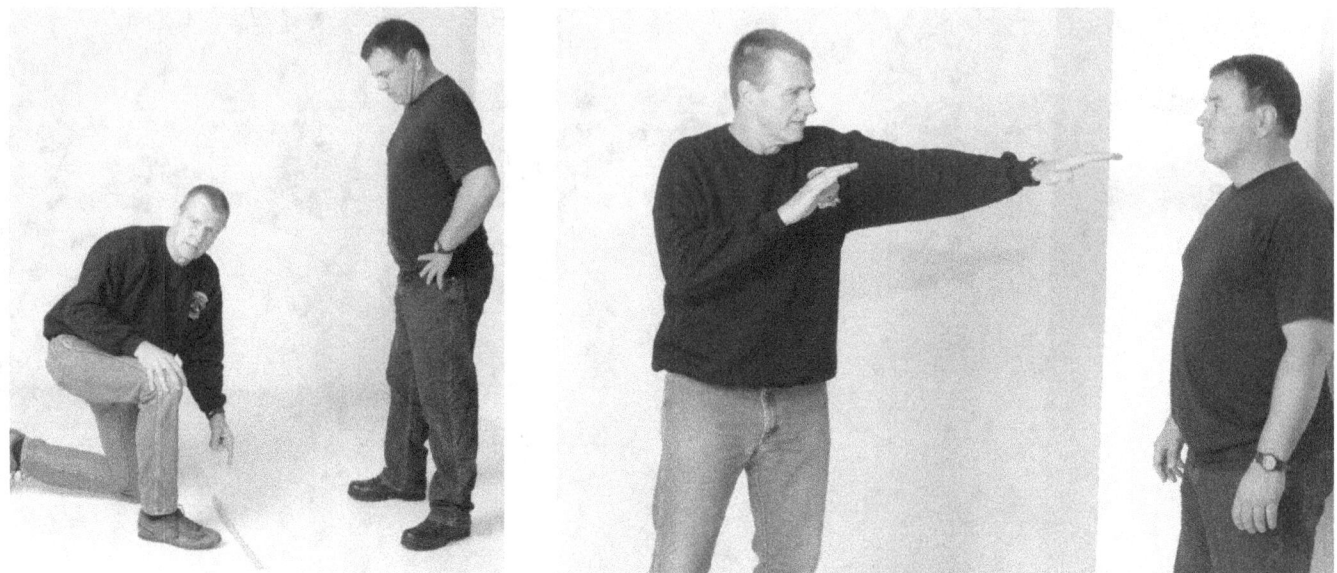

First, use a stick or some device to set a boundary that the trainer should not step past. The trainee stands with his back against the wall. He can only use limbs to block.

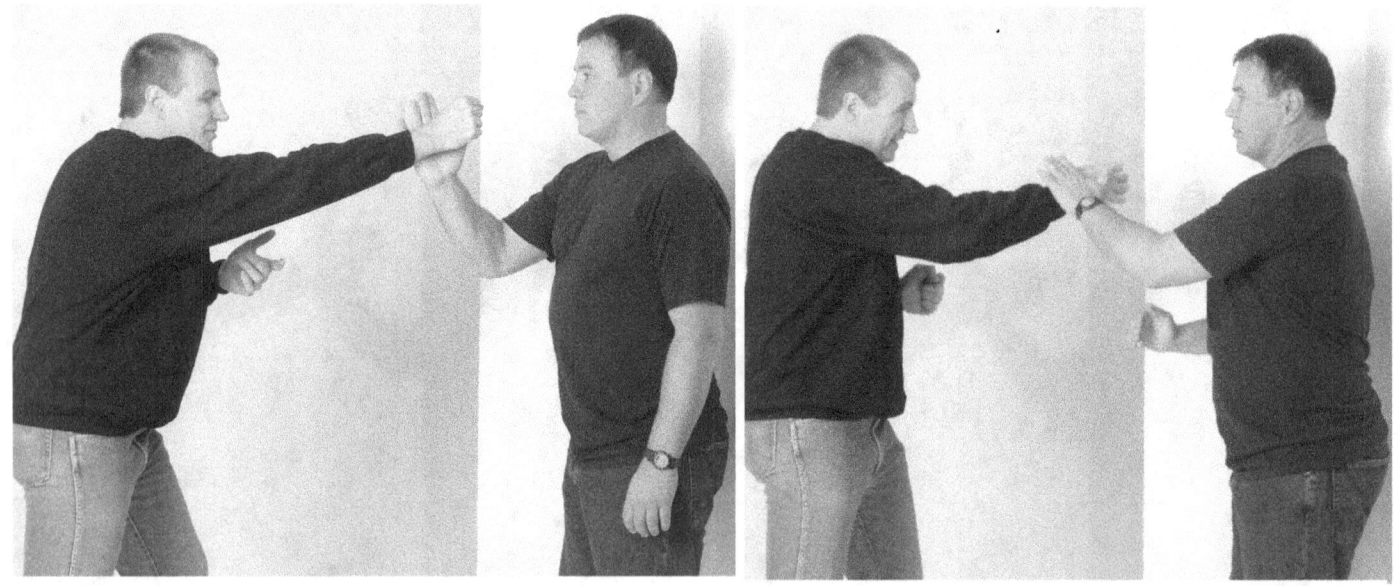

Frontal Version

First, use a stick or some device to set a boundary that the trainer should not step past. The trainee stands with his back against the wall. He can only use limbs to block. The trainer will fire a barrage of hand strikes at the trainee, as close and as fast as possible. He cannot back up. He can only use his limbs to stop and/or deflect the attack.

Bladed Right Lead Version

The trainer sets up at a safe distance to challenge him to his right side. This is more like a bladed fighting position for him, as his body is bladed/angled to one side.

Bladed Left Lead Version

The trainer sets up at a safe distance to challenge him to his left side. This is more like a bladed fighting position for him, as his body is bladed/angled to the other side.

Words and photos cannot describe how fast the trainer attacks and how valuable this drill is.

The Wall Drill Summaries

Trainer takes a safe distance position.

Trainee is backed up against a wall so he cannot back up. He must rely on his blocks.

Trainer strikes out as fast and as furious as possible with:
- eye attacks
- palm strikes
- forearms
- hammer fists
- jabs and crosses
- hooks
- uppercuts
- overhands

Trainee blocks with:
- hands
- forearms
- elbows

Trainer stand before the trainee
- straight in front of him
- off the left side
- off to the right side

Develop more drills from this situation. Use this format as a start and inch inward. Require the trainee to use more skills, counters and follow-ups.

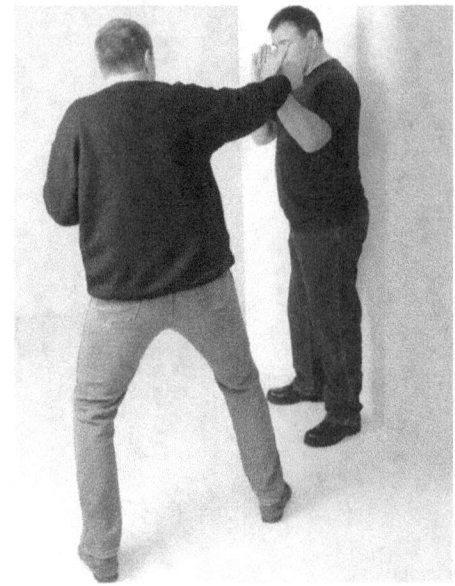

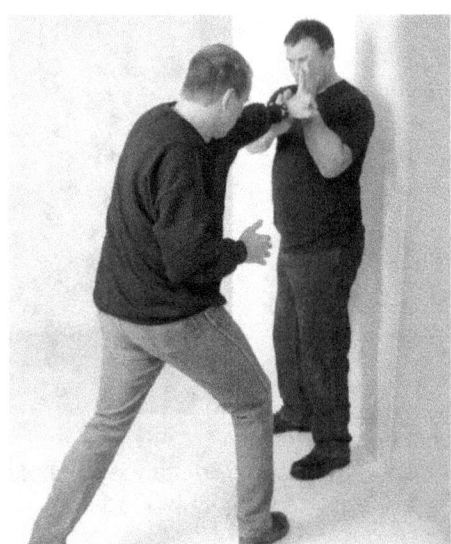

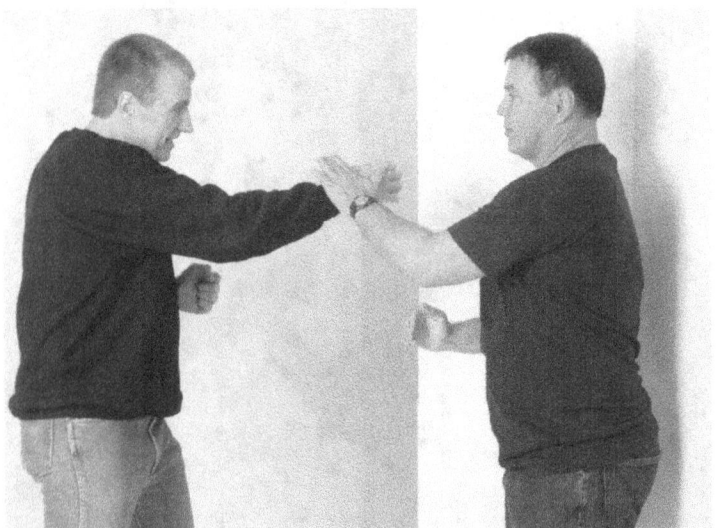

The Statue Drill

By now, readers of the *Training Mission* books know I like to include the statue drill in every module that I possibly can. This introduces new people to body contact in a highly controlled manner. This is not unlike a classic boxing drill called Charlito's Drill. Ducking and weaving is also developed. Pick a punch and work across the body, outside, inside, split, inside, outside.

The Statue Drill
- Outside his right arm (could include his whole back)
- Inside his right arm
- Split between his arms
- Inside his left arm
- Outside his right arm (could include his whole back)

The Straight/Horizontal Blast Drill

This is a very versatile drill with many half-beat insertions. There is a four beat version and a two beat version. The four beat version is really just a two-hand circular style block with a finishing punch. Once this pattern is mastered, you start adding half-beat inserts. Then you work inside the punch. Both arms. Great for speed.

4 Beat Drill

Beat 1: Cross Palm Block
In this sequence, the left hand crosses over.

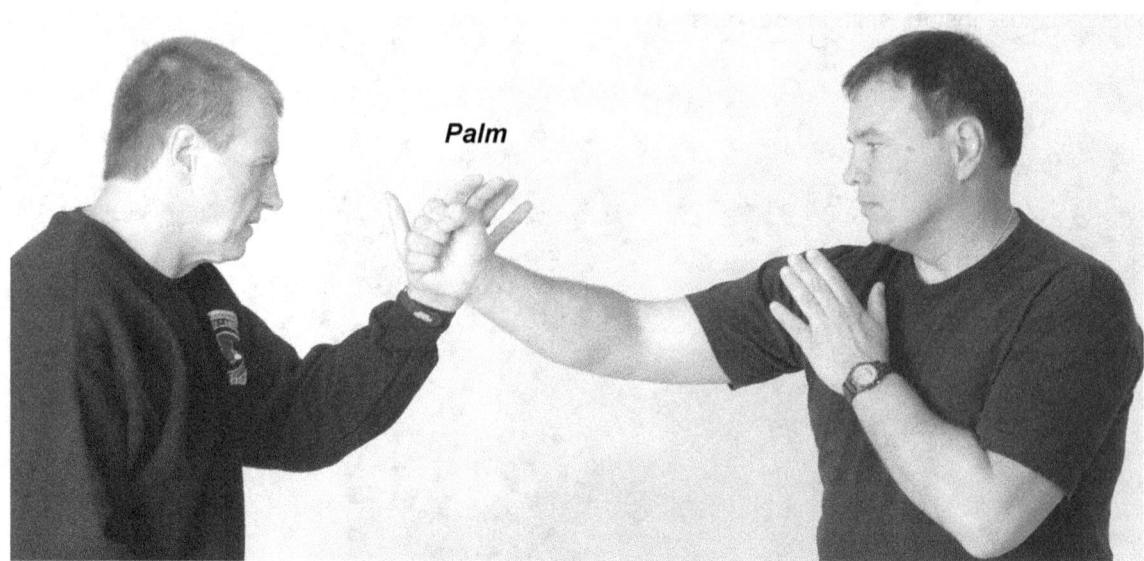

Beat 2: Back Hand Block
In this sequence, the right hand back hand blocks.

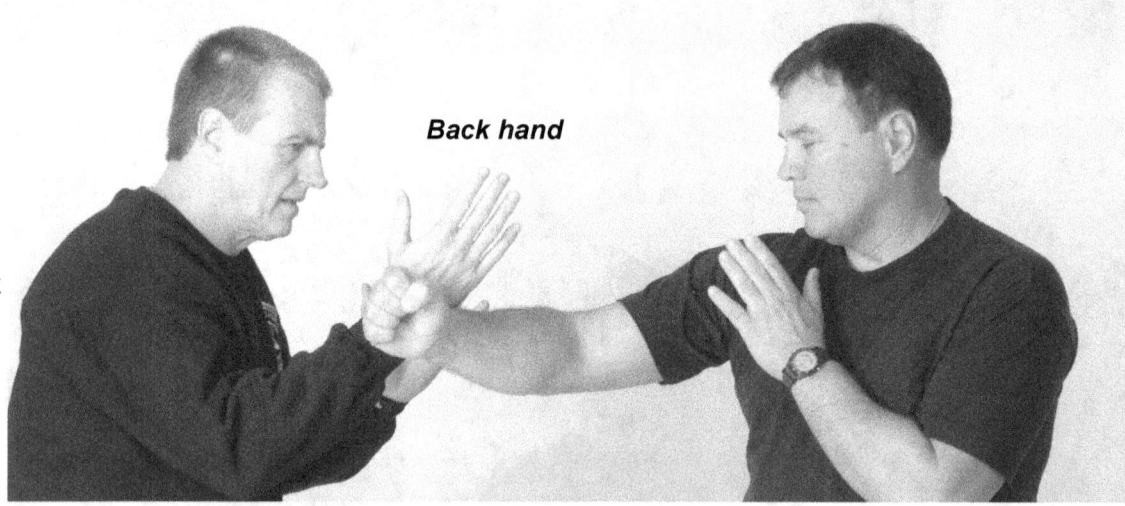

Beat 3: Palm blocks
In this sequence, the left hand blocks.

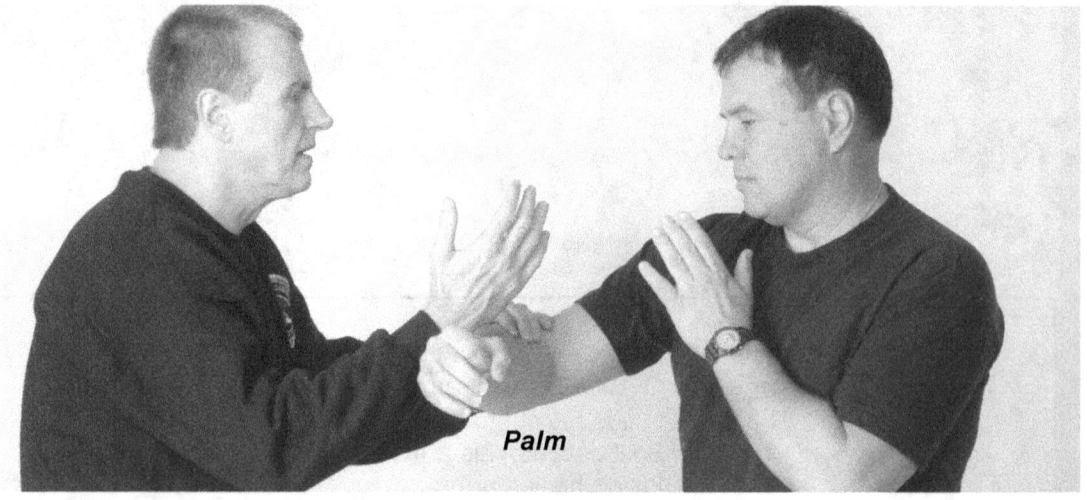

Beat 4: Punch
In this sequence, the right hand punches. The partner begins his own sequence.

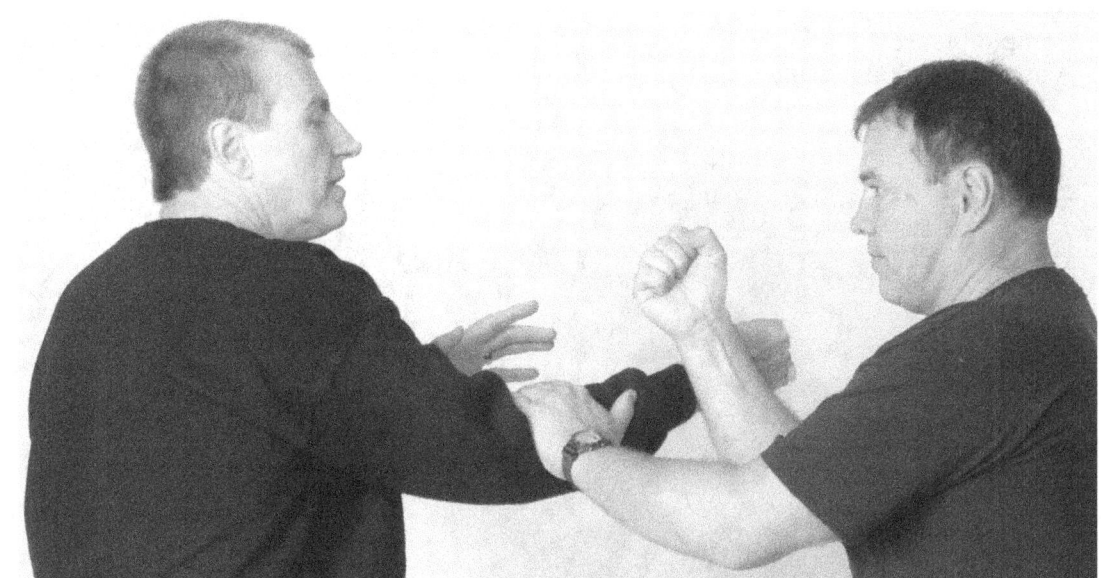

2 Beat Drill

Get the four count started, then:

Beat 1:
Your punch returns with the partner's punch. This looks like a back hand block. Your other hand palm strikes/blocks the punch.

As his punch is blocked, you immediately punch back.

Beat 2:
The partner does the same.

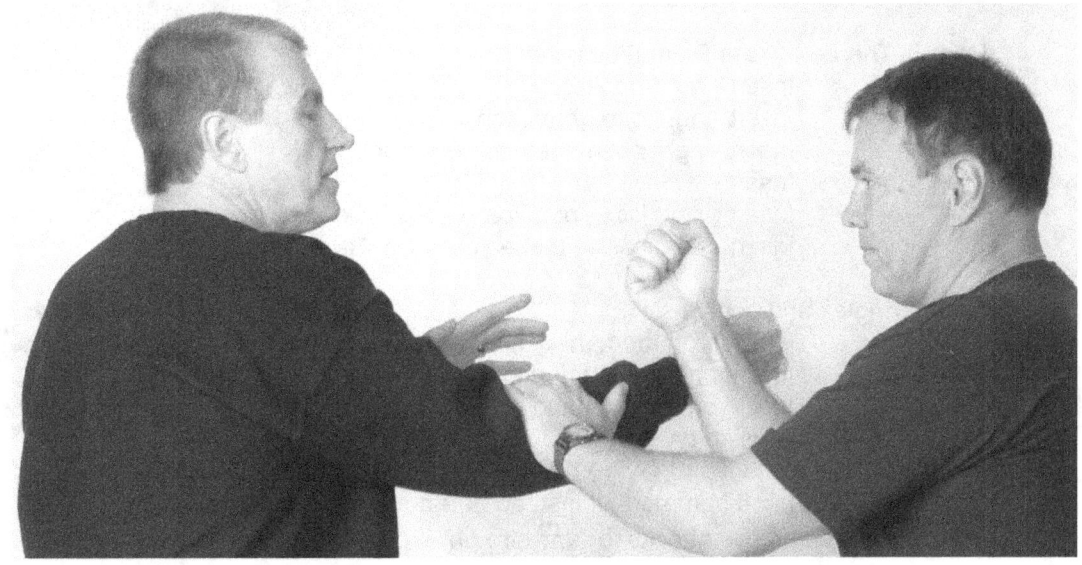

Sample Insert:
A uppercut on Beat 1 1/2

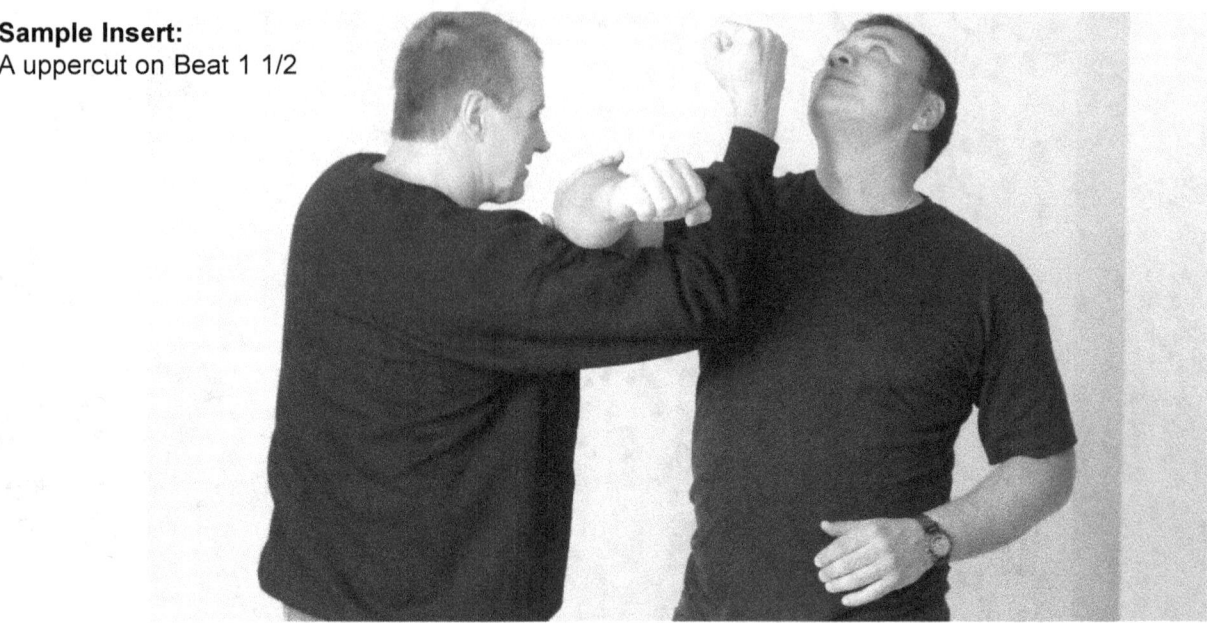

Beat 1 1/2 Samples (too many to list here)

    Beat 1: The Left Cross Palm Block, then...
        Insert - right uppercut
        Insert - right straight punch
        Insert - right palm heel strike to chin
        Insert - left eye jab
        Insert - groin strike
        Insert - shin kick, foot stomp...any kick
        Insert - continue to make your own list

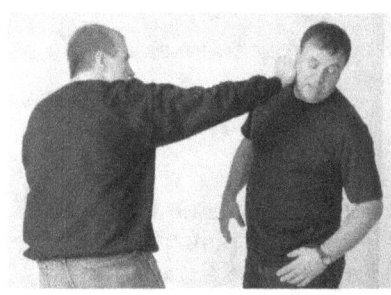

    Beat 2: The Right Back Hand Block, then...
        Insert - left uppercut
        Insert - right straight punch
        Insert - right palm heel strike to chin
        Insert - right eye jab
        Insert - left eye jab
        Insert - right hand grabs and pulls the attack. Strike
        Insert - shin kick, foot stomp...any kick
        Insert - continue to make your own list

    Beat 3: The Left Cross Palm Block, then...
        Insert - right uppercut
        Insert - right straight punch
        Insert - right palm heel strike to chin
        Insert - left eye jab
        Insert - shin kick, foot stomp...any kick
        Insert - continue to make your own list

    Switch Angles and punches
        Outside on the right punch
        Inside on the right punch
        Inside on the left punch
        Outside on the left punch

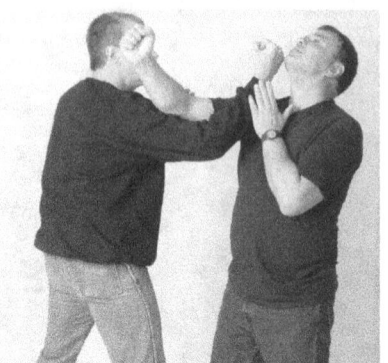

    Practice a back hand hammer fist instead of a punch (a famous drill)
    Practice all versus eye attacks, or any straight arm attack.

Hook Punch Block, Pass and Pin Drill

Since the straight blast drill covered punches coming in straight, we will use only a hook punch in our block, pass and pin drill application. This six-beat format is a staple of our skill and flow drill studies, presented in all prior *Training Mission* books and DVDs.

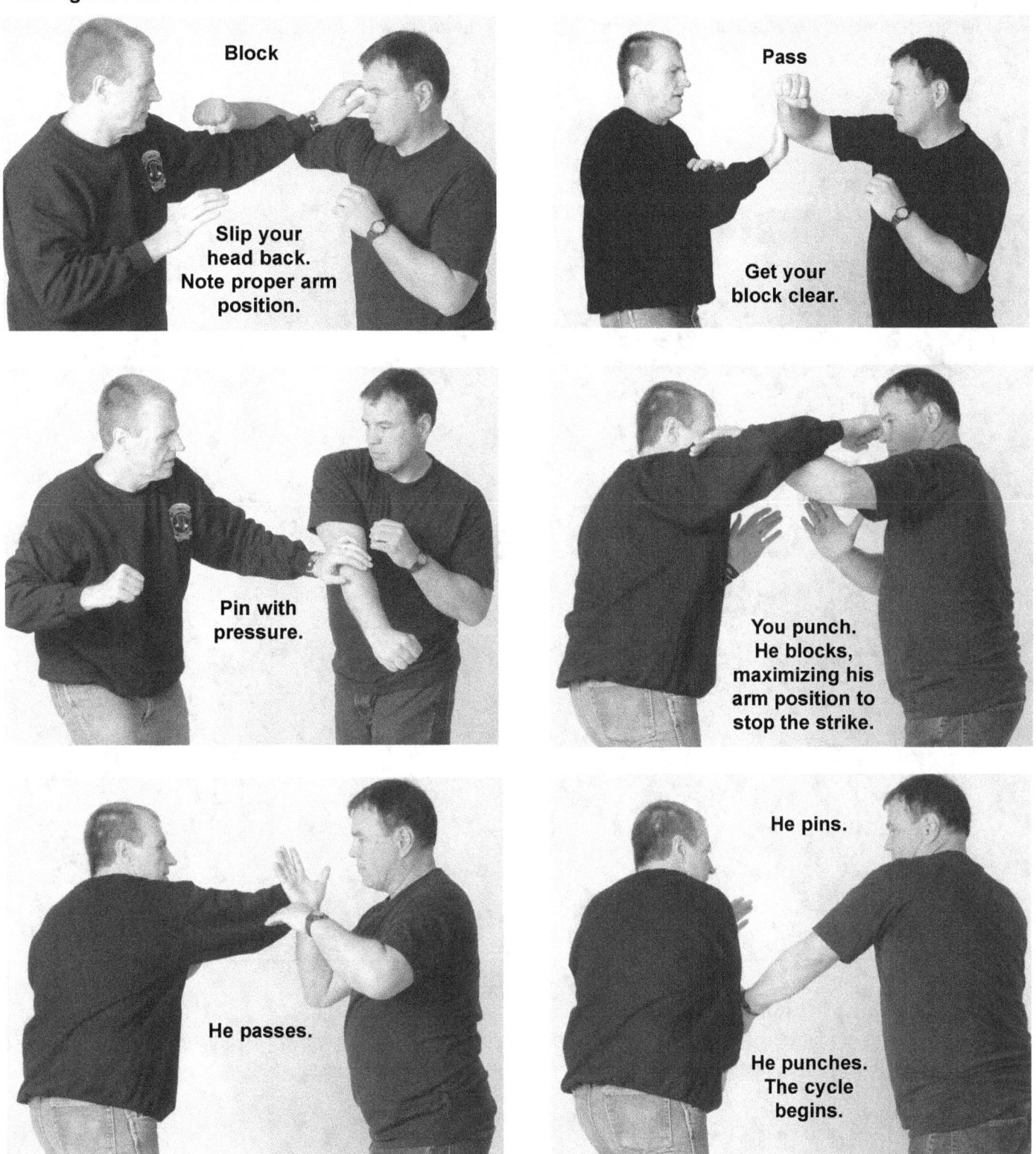

Many repetitions countering the hook punch prepare your reflexes. Make sure you slip your head away from the punch and get a good, partially bent arm block. Work right and left sides. Add half beat inserts, as listed in the straight blast drill. Add half-beat inserts like straight punches, eye jabs, even kicks.

The Sucker Punch

The sudden and sneaky sucker punch is listed by law enforcement authorities worldwide as one of the major tricks with which citizens and police are hit and taken down. Any bit if subterfuge to conceal the punch sets up the victim for the surprise attack. I have demonstrated some common ones here. Pick up on the basics and study them from both the attacker and the victim's positions. Learn how to do it. Learn how to defend against it.

The Common Punching Takedowns

Police training organizations like Caliber Press and criminal justice universities have declared in studies that we are punched and taken down in three ways.

1) The sucker punch.
2) The round-house, haymaker punch.
3) The jab and cross -listed last because most people do not know how to deliver proper jabs and crosses.

Sucker Punch 1) The Basic Turn-Away/Turn Back

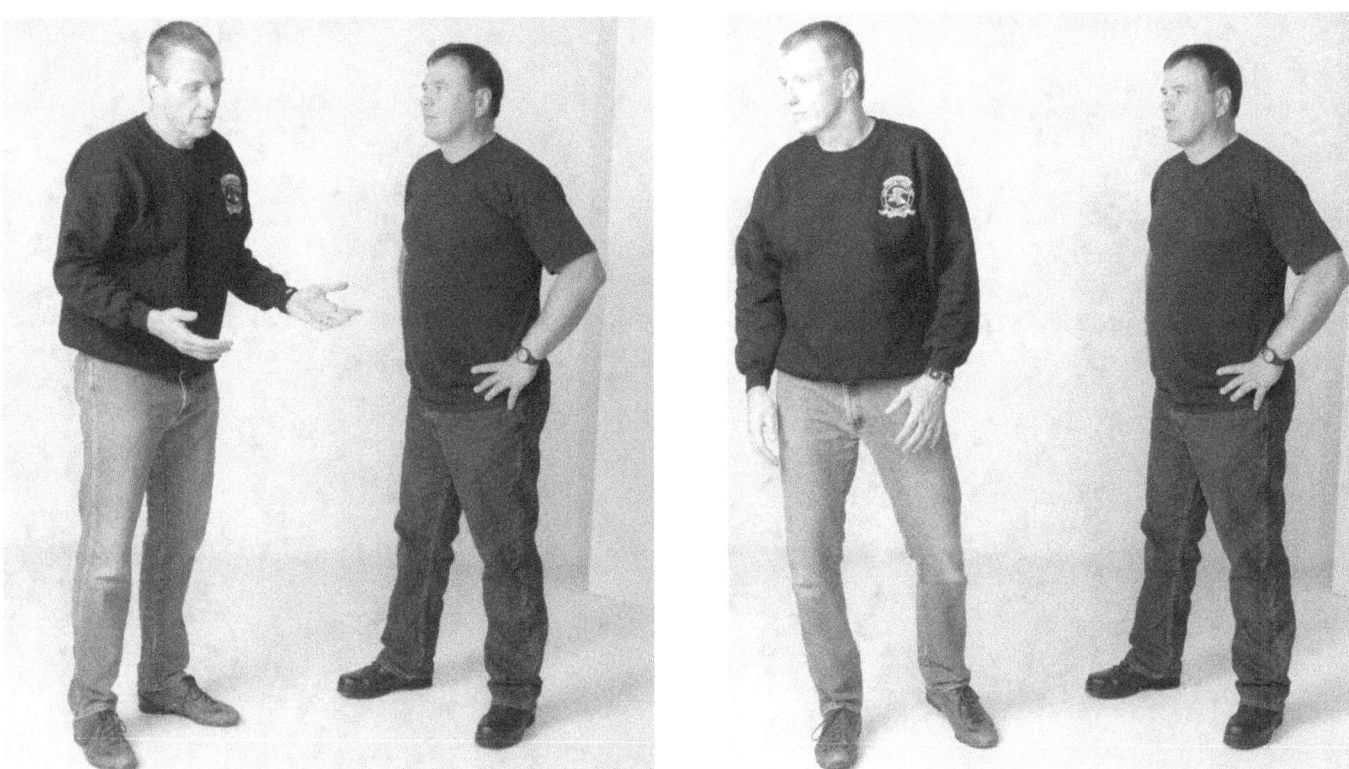

The actor looks like he has calmed down and is leaving. The victim lets down his guard.

With as much concealment as possible, as much power as possible, the actor pivots and strikes a key target.

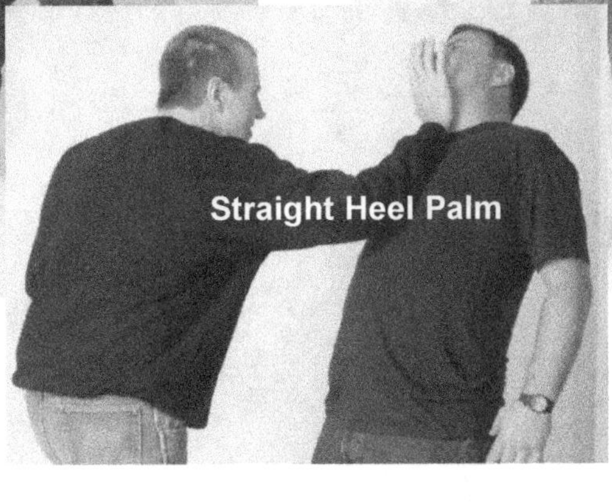

Punch the jaw-line or heel palm strike to lessen the risk of hand injury.

Sucker Punch 2) The Basic Turn-Around: Unblocked

The actor turns and as he completes a turns. he builds momentum and speed, concealing his striking arm. Here a hammer-fist or a forearm strike is in the works.

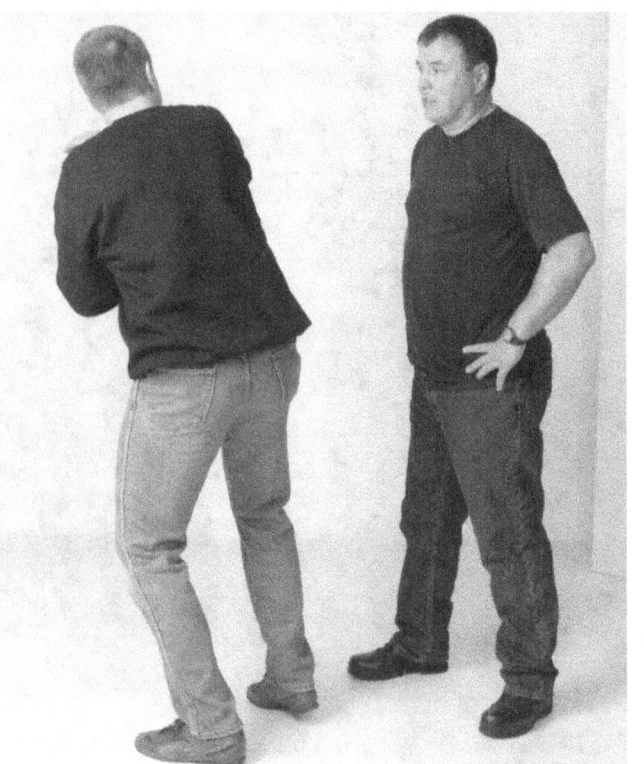
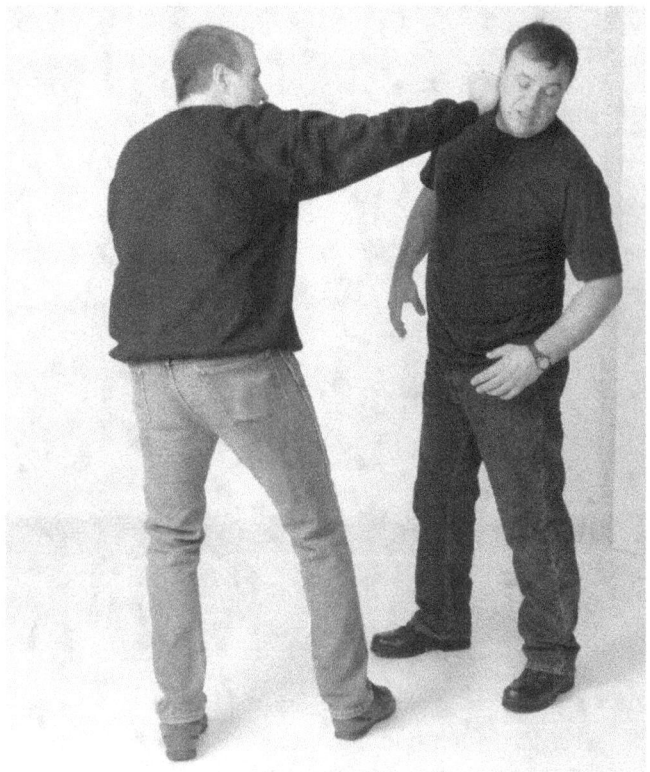

The spin complete, the victim is hit in a key target.

Sucker Punch 3) The Basic Turn-Around: Blocked

The actor turns and completes the spinning circle. The victim reflexively blocks.

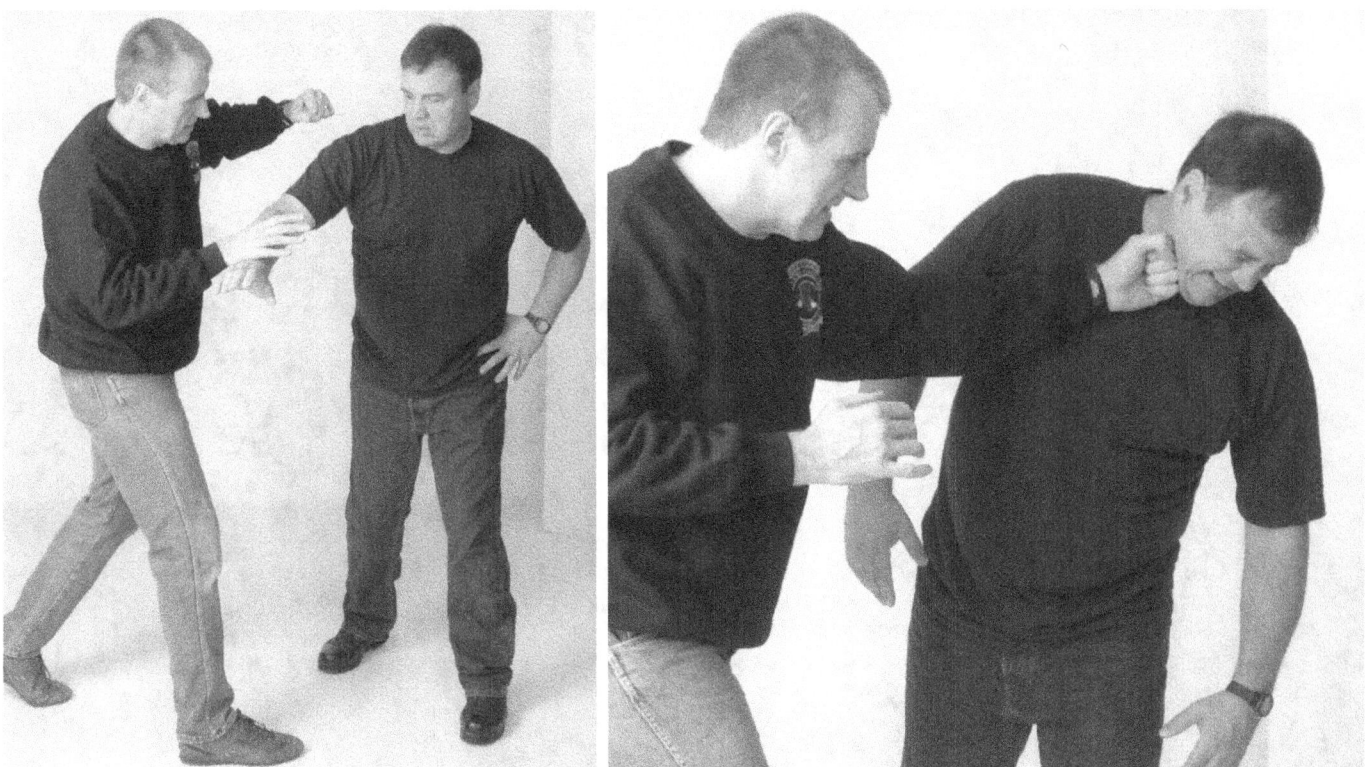

The actor knocks down the blocking arm and follows through with any appropriate strike.

Sucker Punch 4) The Tight Upper Cut

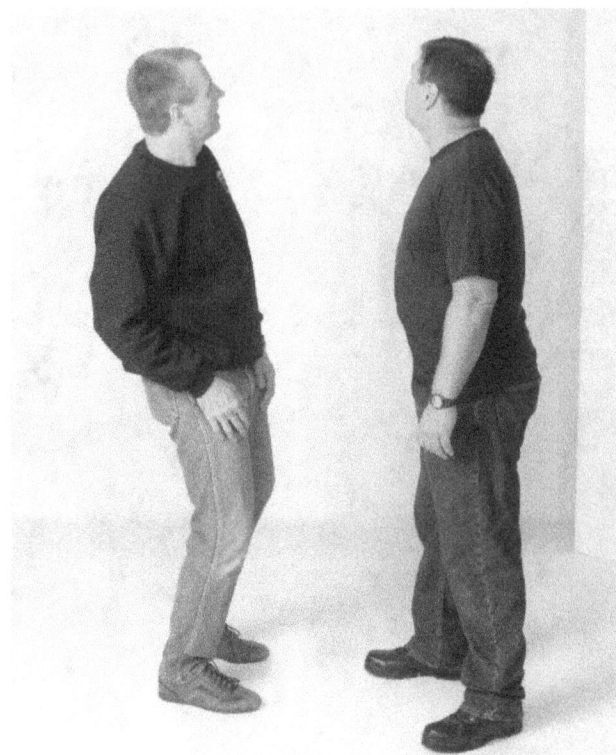

The actor pretends to surrender and calm down. With distracting words, a trick or moves, the actor sets ups...

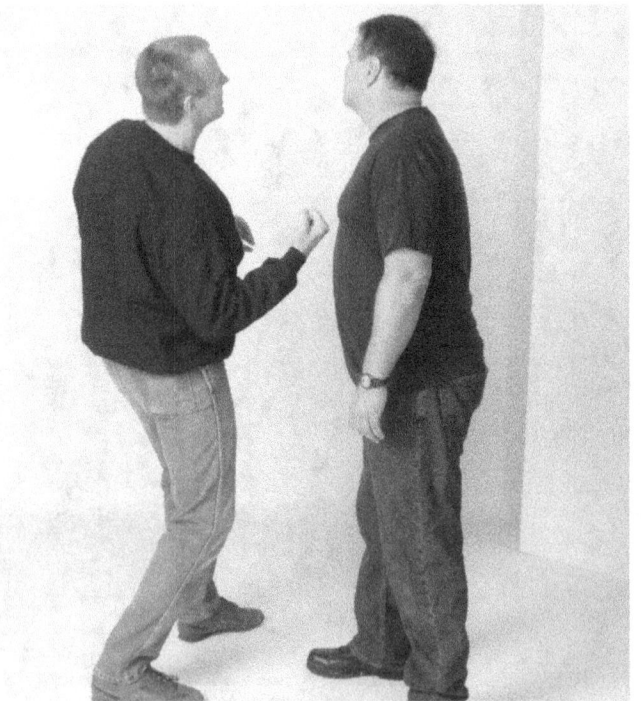
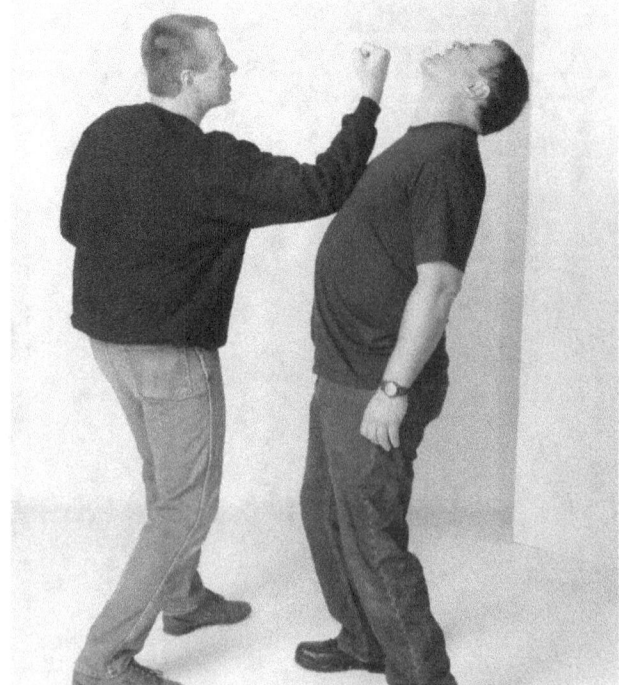

...a tight uppercut, delivered under the chin and under the line of the victim's vision.

Sucker Punch 5) The Distractor

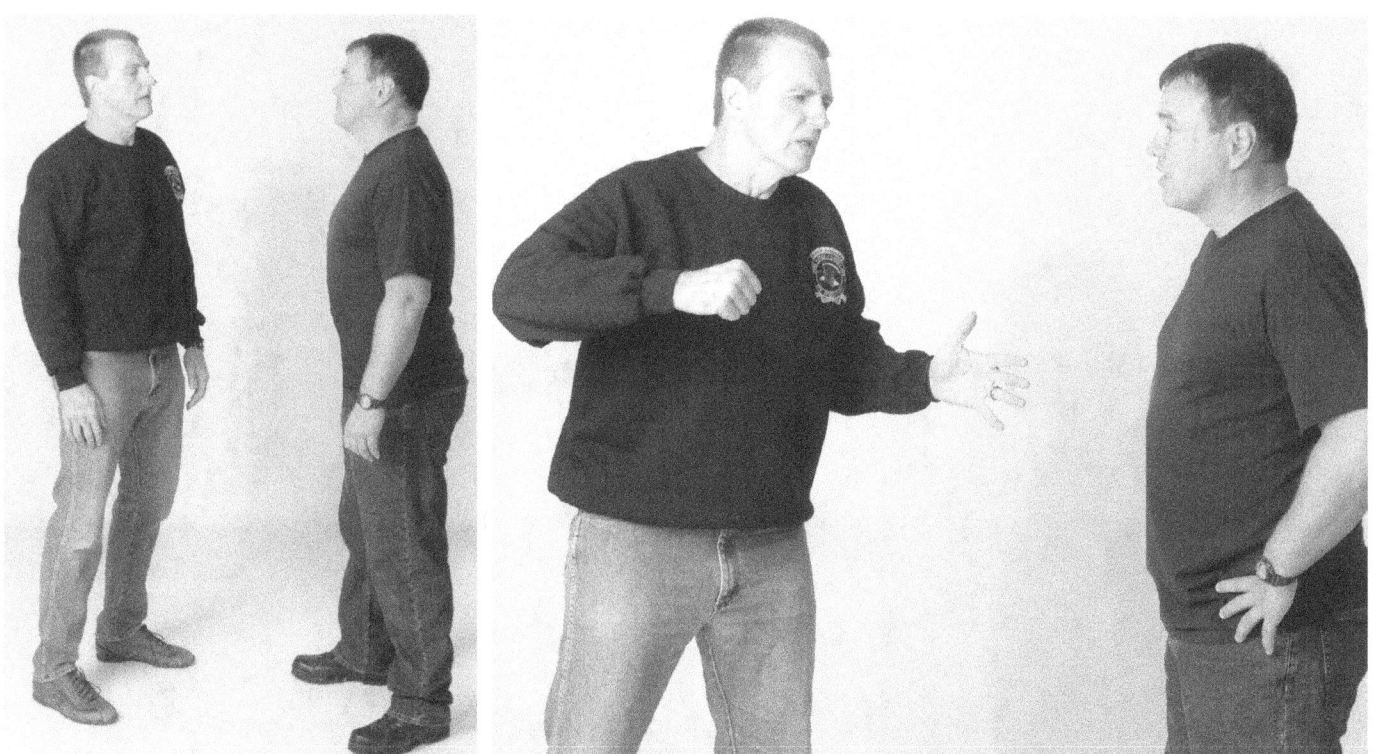

The actor backs down. He raises a fist drawing the attention of the victim. The blindside strike fires in.

Sucker Punch 6) The Handshake

The apology hand shake that sets up a sucker punch.

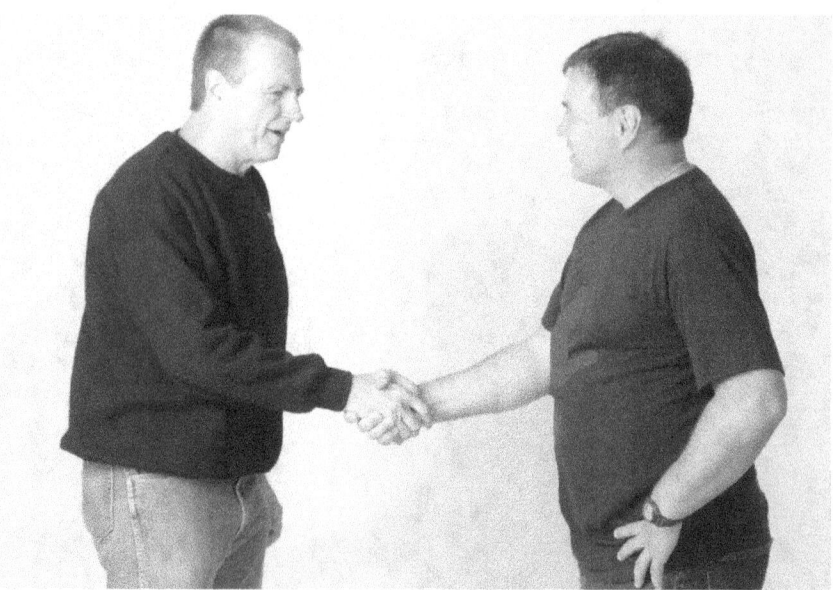

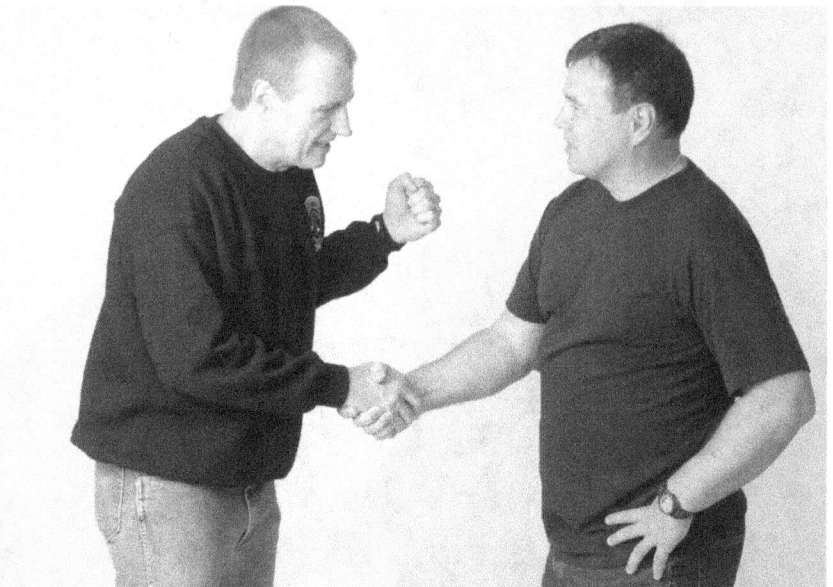

Sucker Punch Targets

* Jaw
* Neck
* Ribs
* Solar Plexus
* Groin

Sucker Punch 7) The Back Hand Slap and Strike

Take an inoffensive stance. Such as one of my favorites shown here. Do not allow your hands to be crossed into your arms and therefore be captured should the opponent charge and pin you.

Back slap the face, opening the neck or jaw.

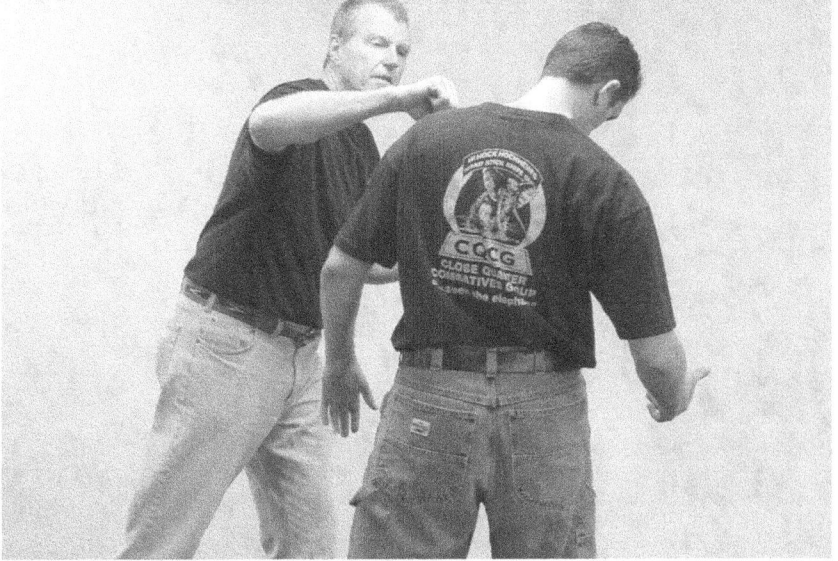

Follow-through with a heat-seeking hook punch to the best opening in the neck. This may require a twisting fist to get the best angle.

Head Escapes: Slips, Ducks, Bobs, Weaves and Rolling with the Punches

You cannot leave the safety of your head and brains to your arms alone. Your head must also help in the dodging process.

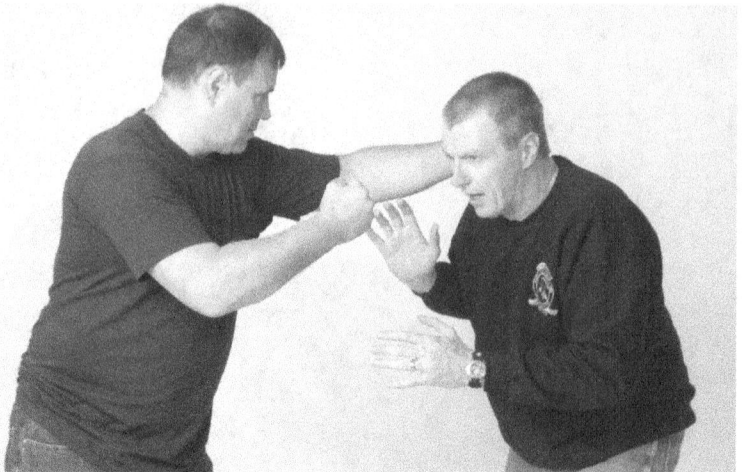

The head drops back.

The head ducks.

The head slips to the side.

Head Dodging Drills

The trainer throws a series of straight punches at your head. You:

    Set 1: dodge back
    Set 2: dodge right
    Set 3: dodge left
    Set 4: duck

The trainer throws a series of hooking punches at your head. You:

    Set 1: dodge back
    Set 2: dodge right
    Set 3: dodge left
    Set 4: duck

The Stunt Man Drills: Rolling with the Punches

If one punch is coming in, one way veterans escape a full punishing blow is to roll with the energy as much as possible.

Always Keep that Jaw Bite Tight!

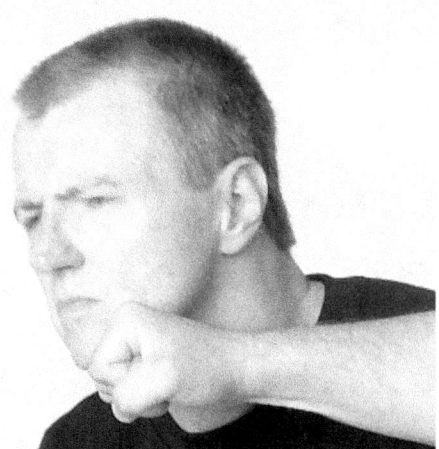

It is imperative that your jaw is locked into position in a fight. A blow to an open, dropped or loose jaw can cause the most debilating, shock provoking breaks a person may experience. The following rehabilitation of a broken jaw is timely, expensive and painful. One good habit created by always wearing a sports mouth piece in practice is it creates a teeth-clenching habit.

The photo series below is self explanatory. You must learn at times to roll with a punch. These steps are classical movie stunt man moves that soften powerful punches. The escapes involve a total body commitment to take the sting from the shot.

People have a natural instinct to take evasive movements against a sudden close-up attack. This fires the body into a dodge maneuver and might develop into a stunt man escape.

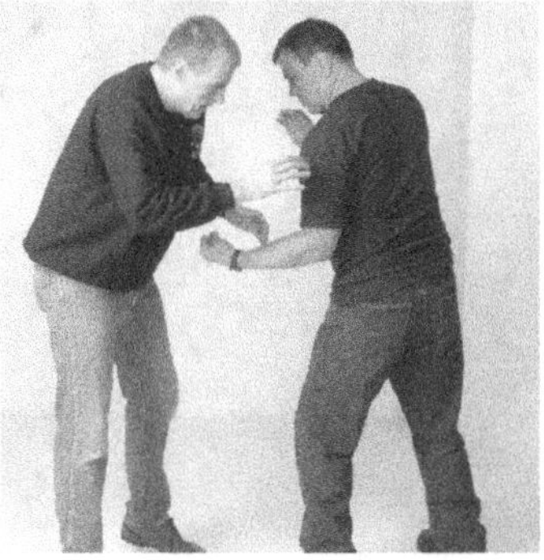

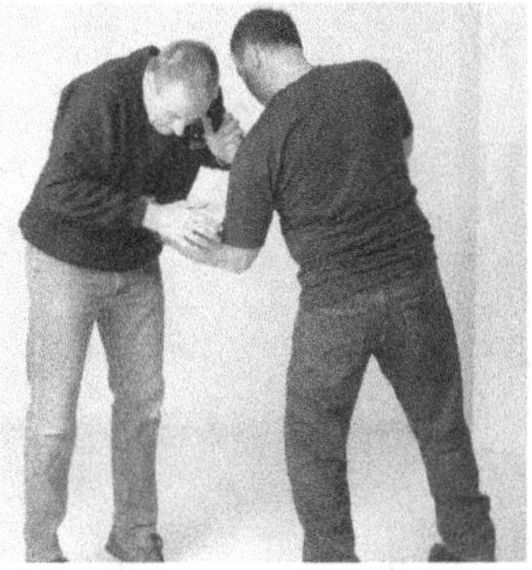

The same strategies work against an uppercut.

Better Punching Through Cables

An old-fashioned cable machine is a great way to develop speed and power in punching. First, your cable machine must be adjustable so you can set the height of pulley. You will set the proper, realistic height for the punch you are exercising. Work the delivery going out and return coming back. This means pushing the weight stack and then also turning your position to pull the weight stack. You must work the push/pull.

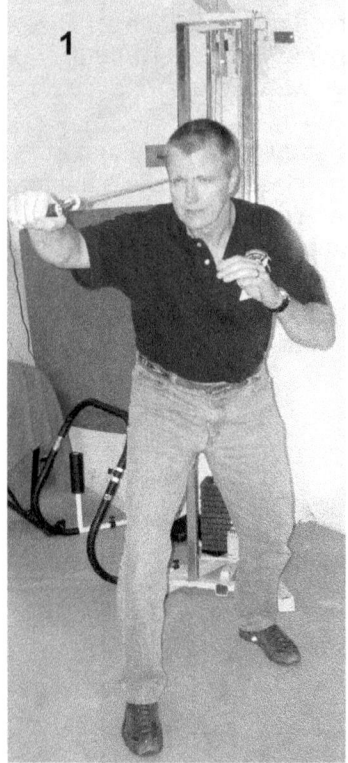
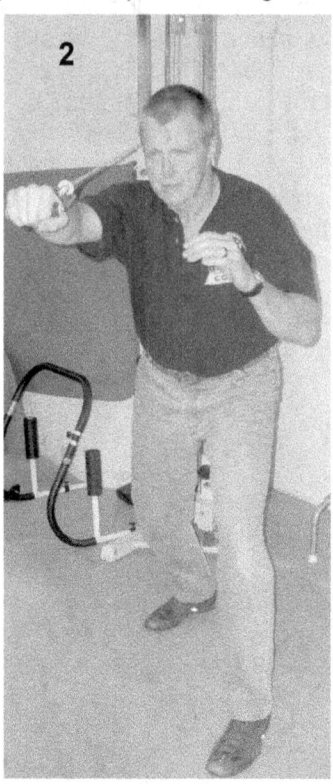

Straight punch right and left in either leg lead. Do sets pulling and pushing the weight. Try to simulate the proper alignment and height. Note in photo 3, I am able to use a corresponding weight rack and handle on the opposite side of this apparatus. This creates a dynamic tension.

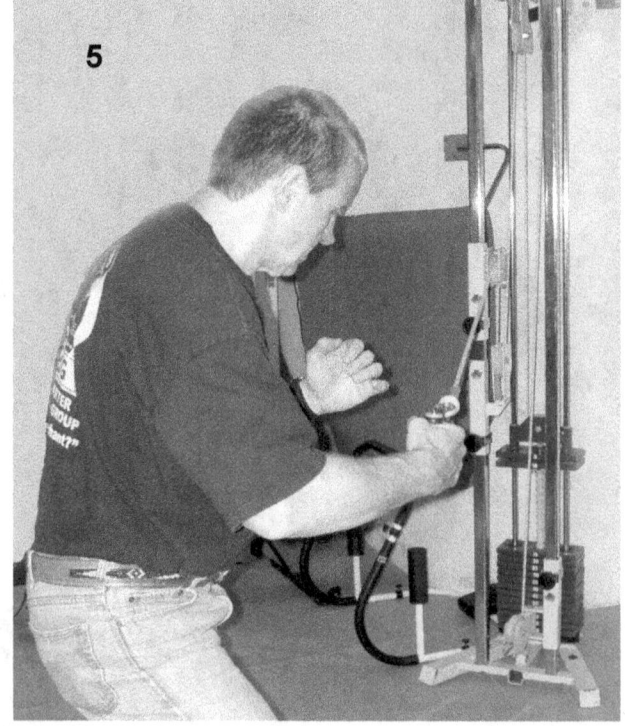

Here I am working both sides of the hook punch. Make sure your body is involved in the exercise.

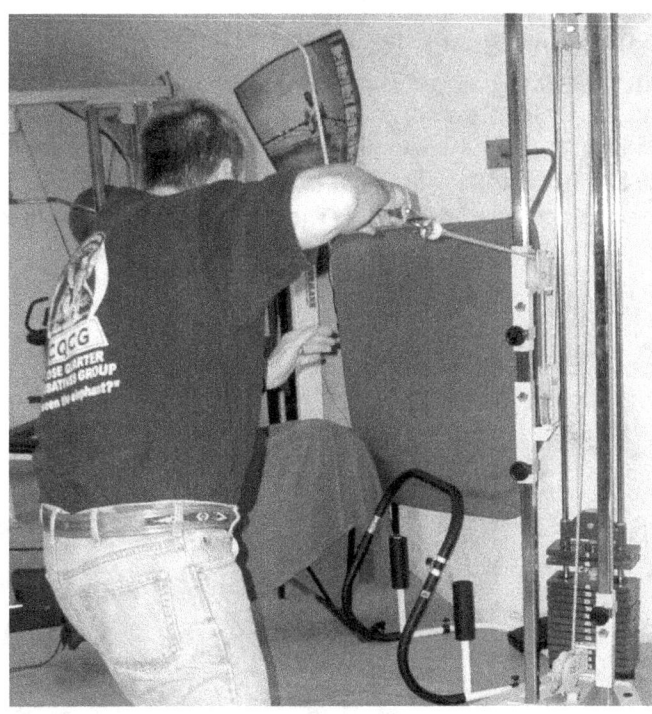

Here are both angles developing the overhand punch. This practice can isolate the intent and focus of the strike.

Second, be very careful of the weight you select. These punch angles and speeds are not conducive to the power-lifting approach. These exercises will not build power-lifter's muscle. Plus, take care not to hyper-extend your joints when punching into the direction of the pulley.

Next, punch faster than the machine's pulley system. What I mean is, you will feel the normal time and distance a cable is supposed to move in the rig with a few reps. Now, when you punch away from the machine, make the weight stack jump with velocity. When the weight drops, punch ahead of the descent. This is why it is so important to make a proper weight stack selection. Do not hyper-extend your elbow or shoulder, or your back.

I have done these cable exercises for 30 years. Even for kicking development, but since your lower back muscles are much more involved in kicks, a lot less weight goes a long way. Hand weights with a topside handle make for strength building tools. You can keep your hands open when you want to practice closed and open hand strikes. You can open your support hand while closing your striking hand. Plus, having this legal tool around makes for a great set of "brass knuckles."

Cable Work-Out

Jab
* back to the machine pulley
* facing the machine pulley

Cross
* back to the machine pulley
* facing the machine pulley

Hook
* back to the machine pulley
* facing the machine pulley

Uppercut
* pulley high
* pulley low

Overhand
* back to the machine pulley
* facing the machine pulley

Hand weights with a topside handle make for strength building tools. You can keep your hands open when you want to practice open hand strikes. Plus, having this legal tool around makes for a great set of "brass knuckles."

The Zone Block Review

In *Training Mission Three* we introduced the basic blocking movements, so I will not reproduce them all here again. Instead I wish to emphasize one very effective, reflexive move, the boxing world often refers to as "zone blocking." You are quite likely to fling your hands up in a zone position as you are placed in more popularized positions. The biggest predictor of what blocking coverage you will snap into, is the position of your arms right before you need the block. The zone, higher arm and lower arm coverage is quite reflexive and a staple in the punch/counter-punch training world.

The ready, hand-up position.

The 12 o'clock zone block.

The 3, 6 and 9 o'clock zone coverage positions. Either arm could be high or low.

The Zone Block - a scenario sample.

The opponent throws a punching attack against your rising or your ready hands.

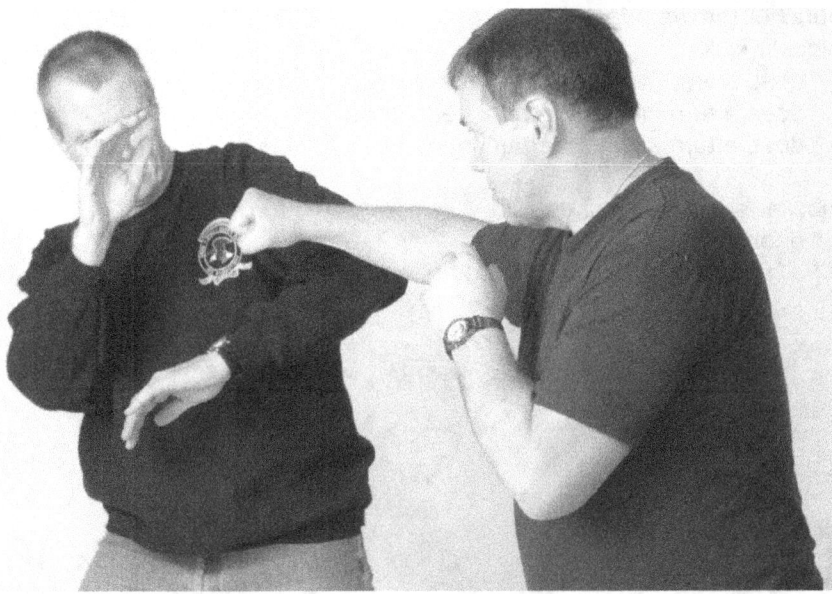

You cover against it in a very reflexive manner, a 9 o'clock position.

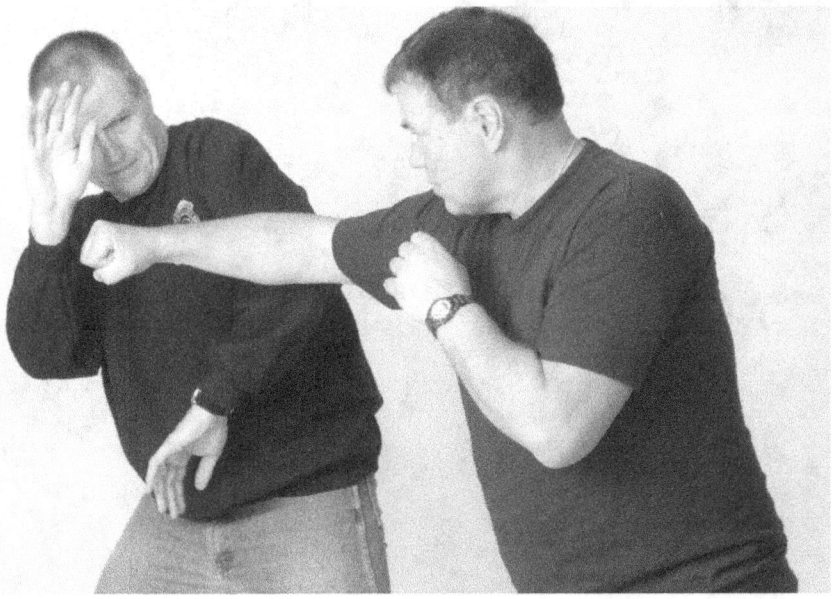

For boxers and boxing trained people, a very common second attack would be a hook punch into your mid section. This zone coverage almost instinctively knows this and offers both a high and low limb in position to go on the defensive or the offensive.

Zone Block/Cross Punch Drill

This is one of my favorite punching drills and overall work-out patterns. I learned the format from Panantukan (Filipino boxing) and Pananjakman (Filipino kicking). I started to build upon the format.

The 4 Beat Basic Training Format

Beat 1) You cross punch your partner's shoulder. He turns and zone blocks, with a good high shoulder.
Beat 2) He cross punches your shoulder. You turn and zone block, with a good high shoulder.
Beat 3) You cross punch your partner's shoulder. He turns and zone blocks, with a good high shoulder.
Beat 4) He cross punches your shoulder. You turn and zone block, with a good high shoulder.

- Now stalk each other.
- Begin again at Beat 1 for a set of 4.
  (Note: make sure you really blast that shoulder. It will be good for the both of you.)

The Advanced Training Format.

The puncher cross-punches and throws a hook into the body
The puncher cross-punches and throws a second punch (what would be a jab-side punch)
At the end of the 4th beat, the puncher throws a round house kick
At the end of the 4th beat, the puncher throws a lead hook kick
At the end of the 4th beat, the blocker kicks:
* does a lead leg hook kick
* does a turn kick to the puncher's knee
* does a turn kick to the puncher's ribs

At the end of the 4th beat, the blocker spins and strikes with:
* a forearm
* a hammer fist (a back fist could lead to a broken hand)

Continue making your own list.

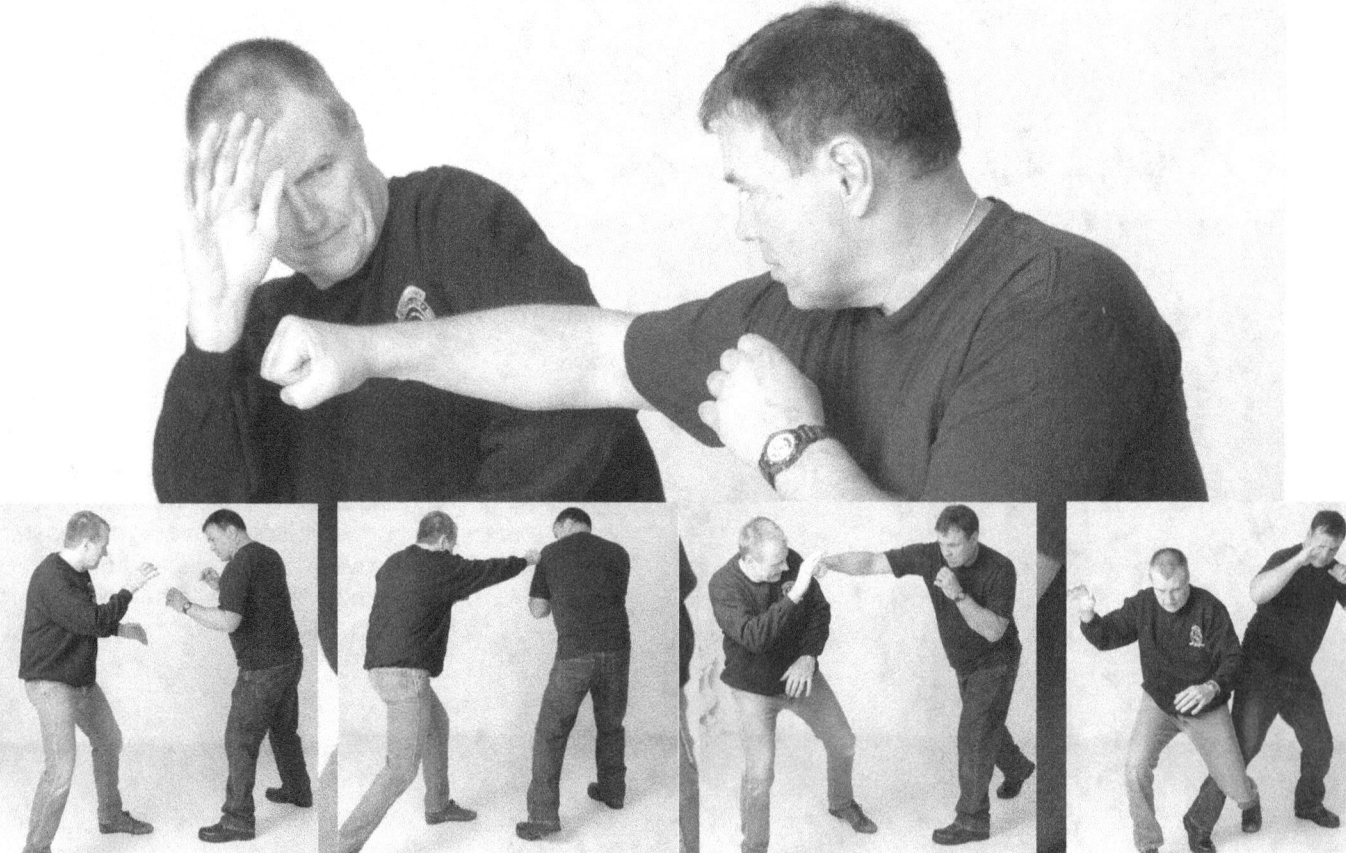

Unarmed Combative Level 5 Kick: The Lead Leg Hook Kick

This is a handy kick that, when used at the appropriate time, can be an effective weapon. It can be executed two ways, inward or outward. The striking surface includes the tip of the shoe, the instep and the shin, but the primary striking surface is the shin. When executing from standing positions, usually the rear foot has to shuffle in or off to the side to support the kick. When executing from a ground fight position, the outward hooking angle is unique. The inward hooking angle resembles the same mechanics of any ground round/hook kick, as we have done in Level 4.

The lead leg, outward hook kick. The shin is delivered on an outward angle, like 10 o'clock on the combat clock.

The left lead leg strikes outward. *The right lead leg strikes outward.*

The left lead right strikes inward.   *The left lead leg strikes inward.*

Your Lead Leg Hook Kick Workout

    Inward Lead Leg Hook Kicks - Standing

        10 hook kicks left lead leg
        10 hook kicks right lead leg

    Outward Lead Leg Hook Kicks - Standing

        10 hook kicks left lead leg
        10 hook kicks right lead leg

    Inward Kicks - Ground

        10 hook kicks left lead leg
        10 hook kicks right lead leg
        - Get up safely at the end of 10
        - Note that these inward
          ground hooks resemble the
          the rear leg ground hook.

    Outward Kicks - Ground

        10 hook kicks left lead leg
        10 hook kicks right lead leg
        Get up safely at the end of 10

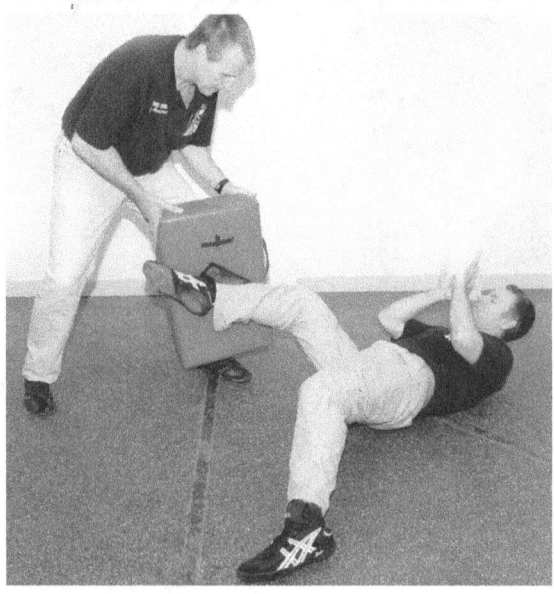

The Level 5 Takedowns: the Frontal Takedowns

This is a basic collection of takedowns that involve forcing the opponent straight down onto his chest and face.

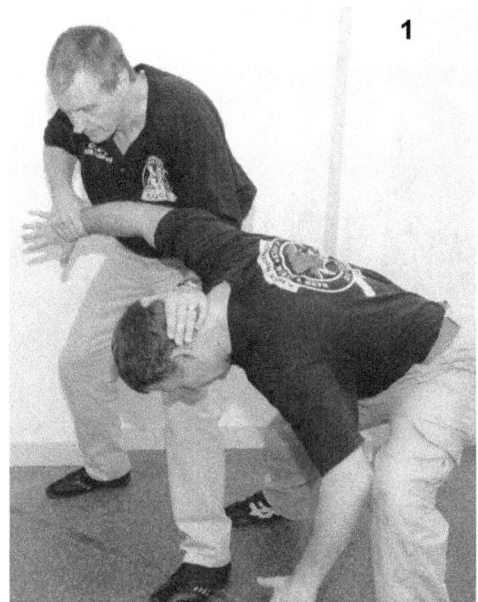

1) The Half-Nelson TKD

2) The Belgium TKD

3) The Face Pull TKD

4) The Counter-Tackle Sprawl TKD

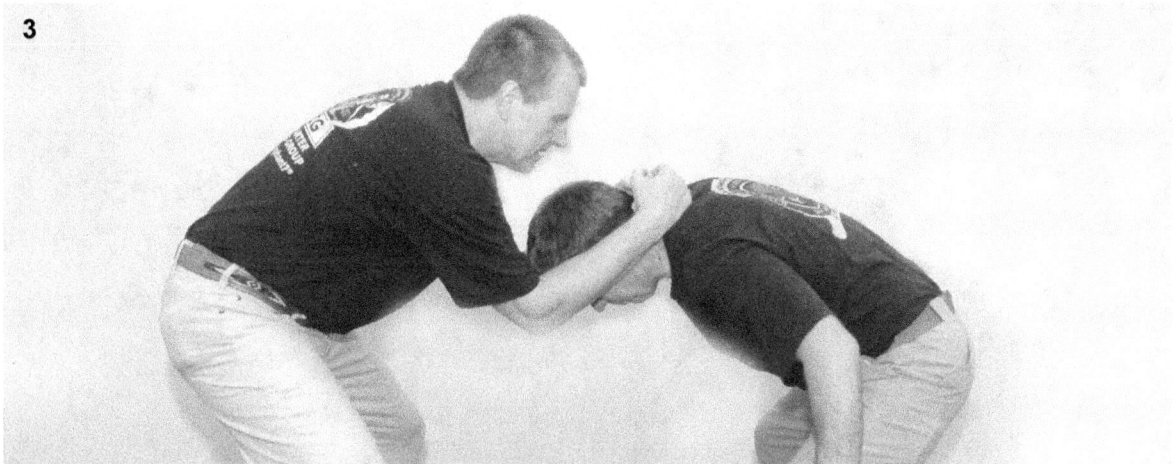

The Frontal Takedowns: The Half Nelson Front Takedown
If the enemy is sufficiently stunned, you can drive him face down in this side-by-side manner.

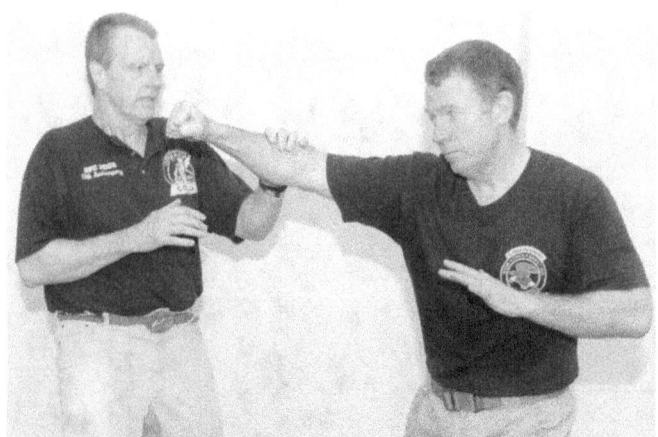

He punches. You dodge.

Grab the hand and elbow - strike the torso as you...

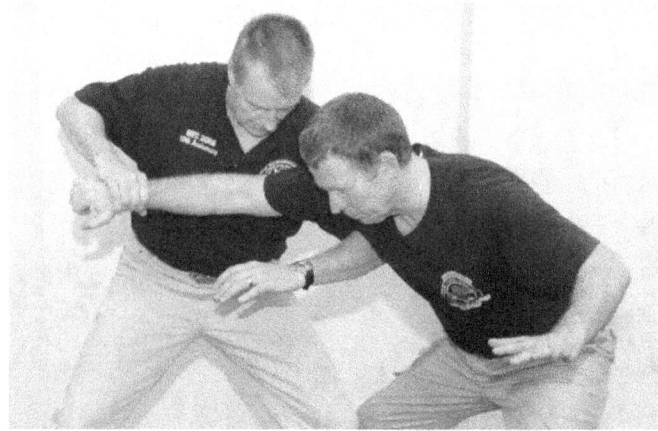

...strike the groin.

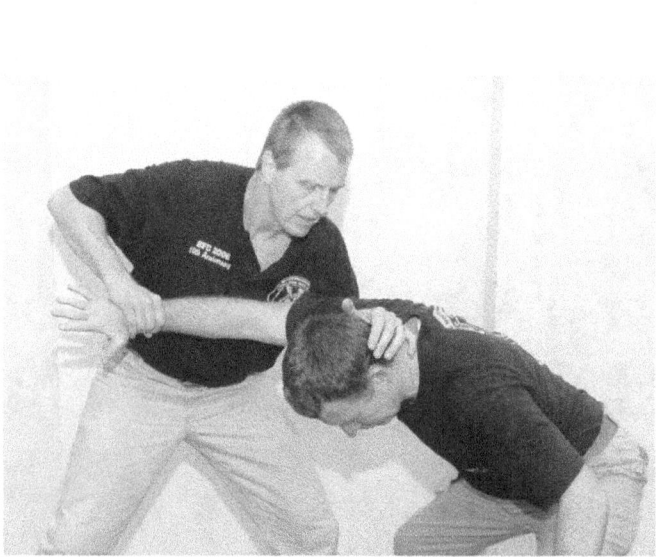

If he bends, hook the hand over the neck as in the classic "half-nelson" capture.

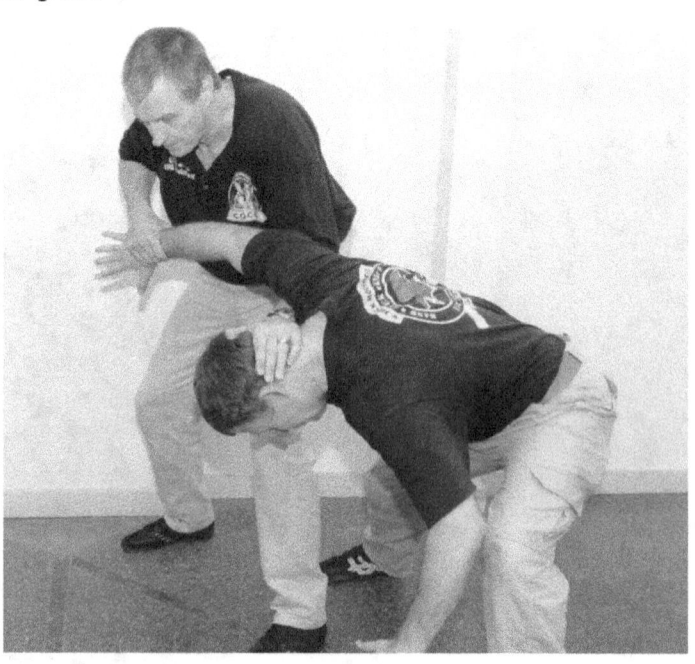

*Note the configuration of the arm under his armpit and over his neck. This is the classic Half-Nelson capture. Circular, wheel throw versions of this appear in **Training Mission 2**.*

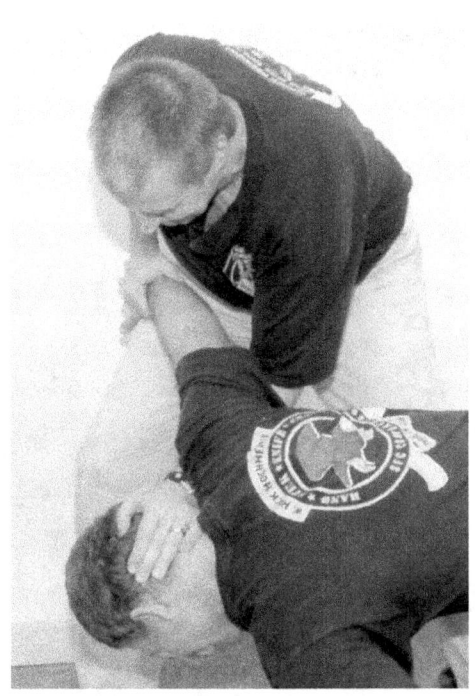

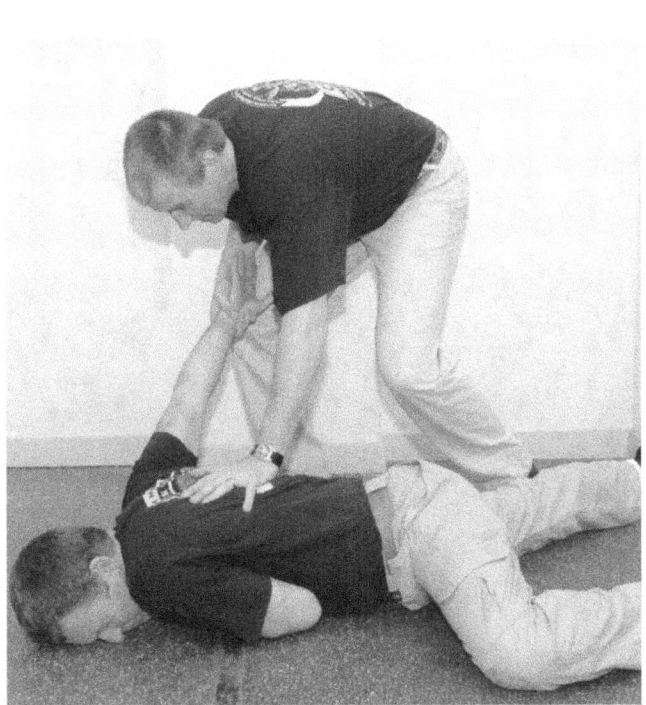
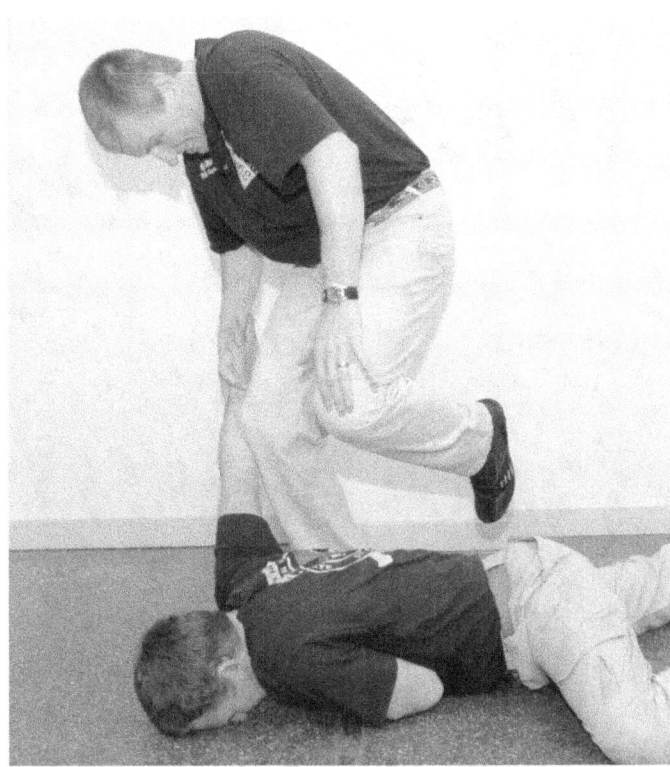

You finish the fight as needed for the situation.

Notes:
The subject may instinctively do a forward roll, which does afford him some measure of relief but does not help him completely escape. Other counters have the captured man diving between the legs to escape the chest-down, face smashing, end capture.

Frontal Takedowns: The Belgium Takedown

I do not know why this is called the Belgium Takedown, but it has been a staple in most military tactics for decades. You creep up on the enemy, hook his ankles, shoulder tackle against his knees while pulling the ankles back with a slight lift to the pull. This was primarily a sentry takedown, but offers no silence in the stratagem, as the opponent can gasp aloud or yell.

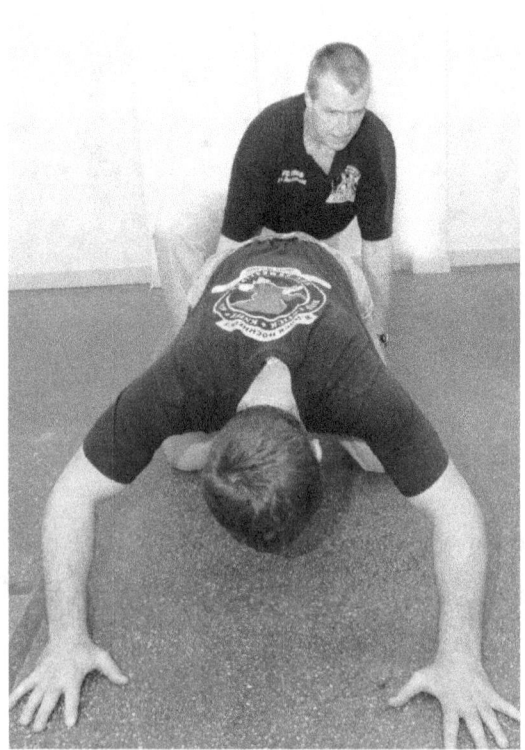

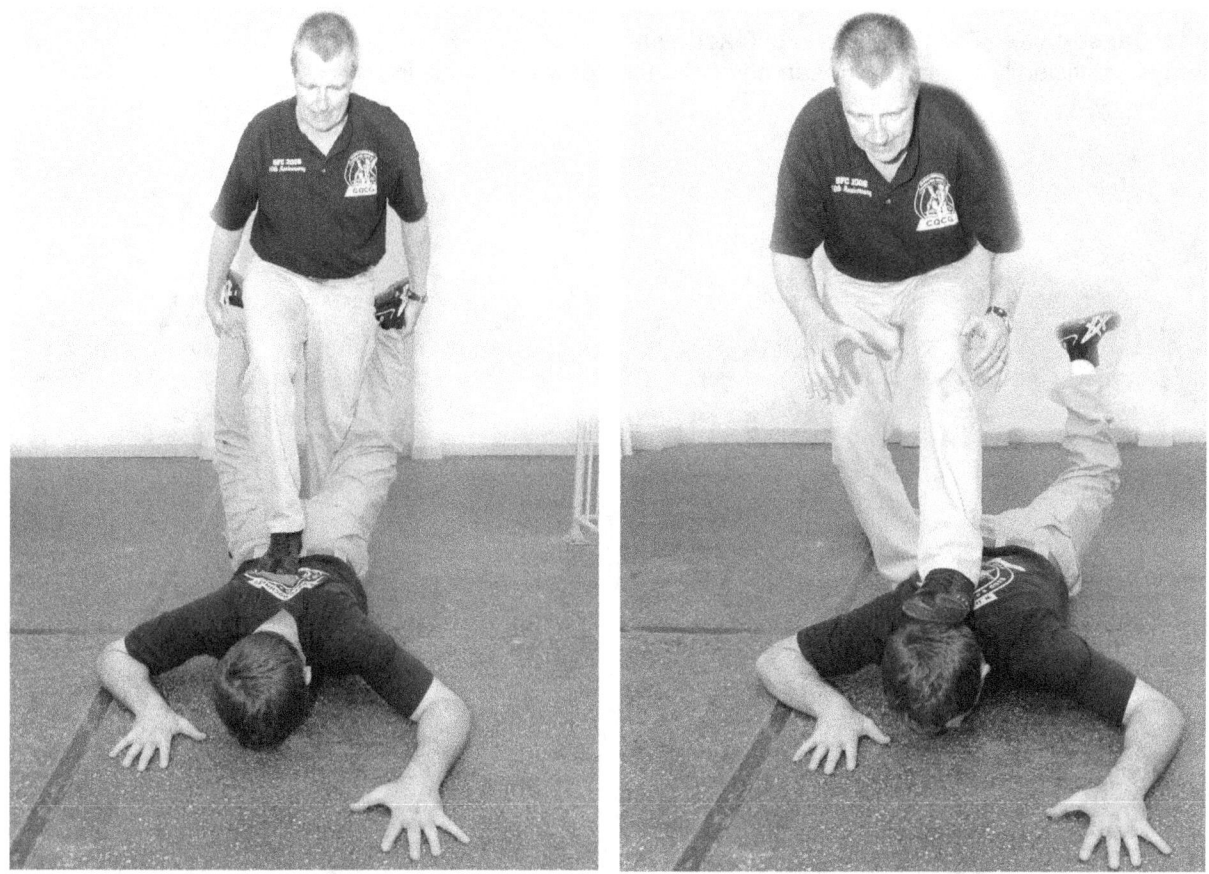

The Reverse Boston Crab, a step away from back-breaking, lethal force.

The finish to this is especially brutal and not taught within most military doctrines. Former French Foreign Legionaire Nick Hughes showed me this back-breaking finish the Legion taught. When the enemy hits the deck, you secure his ankles and start hauling his legs forward. You stomp-step violently on his back. When you get the desired result, you can let loose of the legs and finish the move with a kick. Remember the last kick is onto a rounded object (the head) so take care not to damage your ankle.

Also take extra care when practicing this finish. Note in the photos that I am not gripping Doc Shelton's ankle too tightly so as to give him some freedom of movement. Tightly arm-wrapping the lower legs injures the real enemy. In practice we must take care not to injure the lower back of our training partner.

Note:
There are some escapes from the regular and reverse Boston Crabs. An effective and little known one is done in the mid-to-early phase. The man tries to do a small push-up, creating just enough space to tuck an arm deep across his chest to the far, lower side. He follows with his shoulder until his whole upper torso is turned and the small of his back is now protected from the back-breaking crunch. This position makes it hard for the attacker to grapple with you.

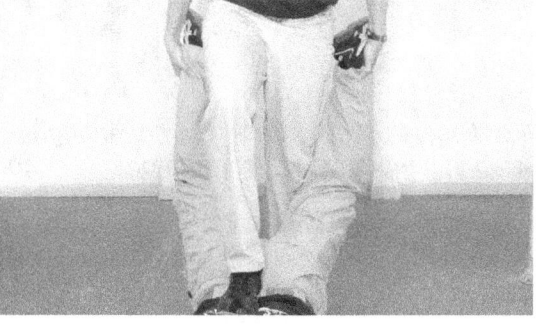

Do not injure your friend's lower back when you practice this move.

Keep your grip loose, and bend loose to save injury.

The Frontal Takedowns: The Front Sweep Takedown
If the enemy is sufficiently stunned, you can drive him face down in this side-by-side manner.

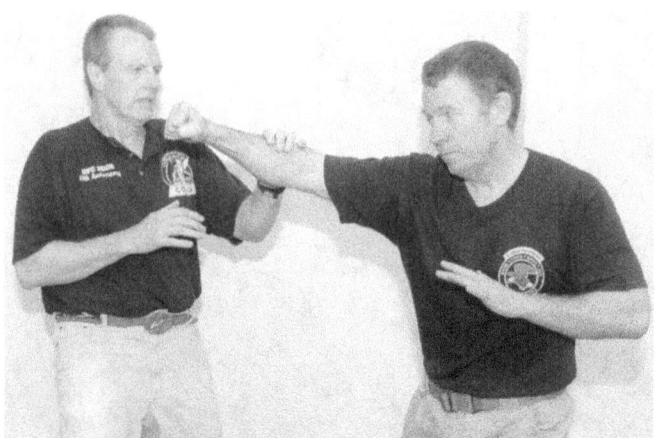
He punches. You dodge.

Grab the limb and elbow strike the torso as you...

...strike the groin. Do what it takes to get the proper bend.

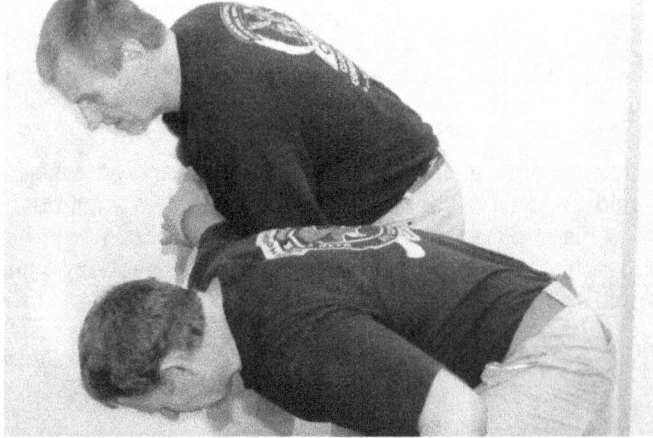

Some systems suggest a quick, fingers-up eye attack, but not enough pressure to cause the head to rocket back up. Then you step in deep and put your arm or lower shoulder into his armpit and...

Sweep his leg.

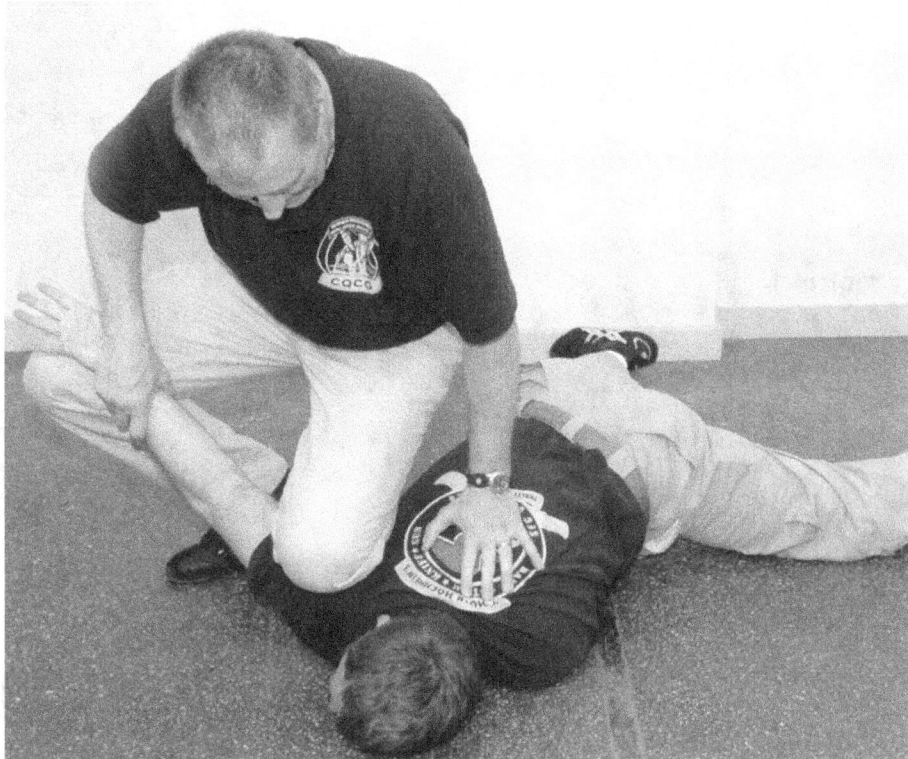

...You sweep his inside leg as you bang into his arm pit for a takedown. This is a capture-finish popular with police officers. His elbow is hyper-extended.

The Three Basic Foot Position Problems
Your opponent's inside foot and leg may be in one of these positions, in relation to your position, while you attempt this sweeping takedown.

> Position 1: His inside foot is far back.
> Position 2: His inside foot is in mid-range.
> Position 3: His inside foot is far forward.

1) Foot far back.

2) Foot in mid-range.

3) Foot too far forward.

Page 71 - W. Hock Hochheim's Training Mission Five

**Problem 1
Foot far back.**

1) Foot far back?

Hook ankle and drive torso sideways for a tripping takedown.

2) Foot far back?

Kick knee sideways.

**Problem 2
Foot mid-range.**

Foot in midrange?

Use the classic foot sweep as you drive the torso forward up on his shoulder.

Page 72 - W. Hock Hochheim's Training Mission Five

1) Foot Forward?

Reach down and hook leg...

Problem 3 Foot Far Forward.

...pull up and back.

2) Foot forward?

Leap into the air and drop a knee on the shoulder. It should wipe almost anyone out.

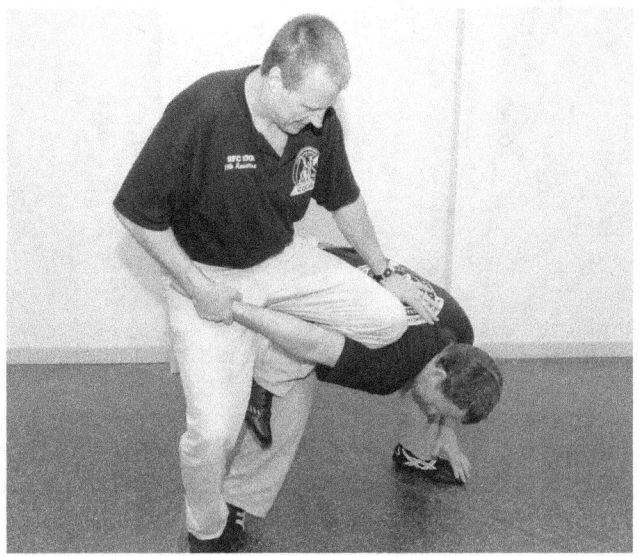

Frontal Takedowns: The Head Pulldown

Simple and effective if done against a stunned opponent. Basically, with a solid grip on his neck and back of the skull, you put his face on the ground. This involves taking a few steps back to clear his fall.

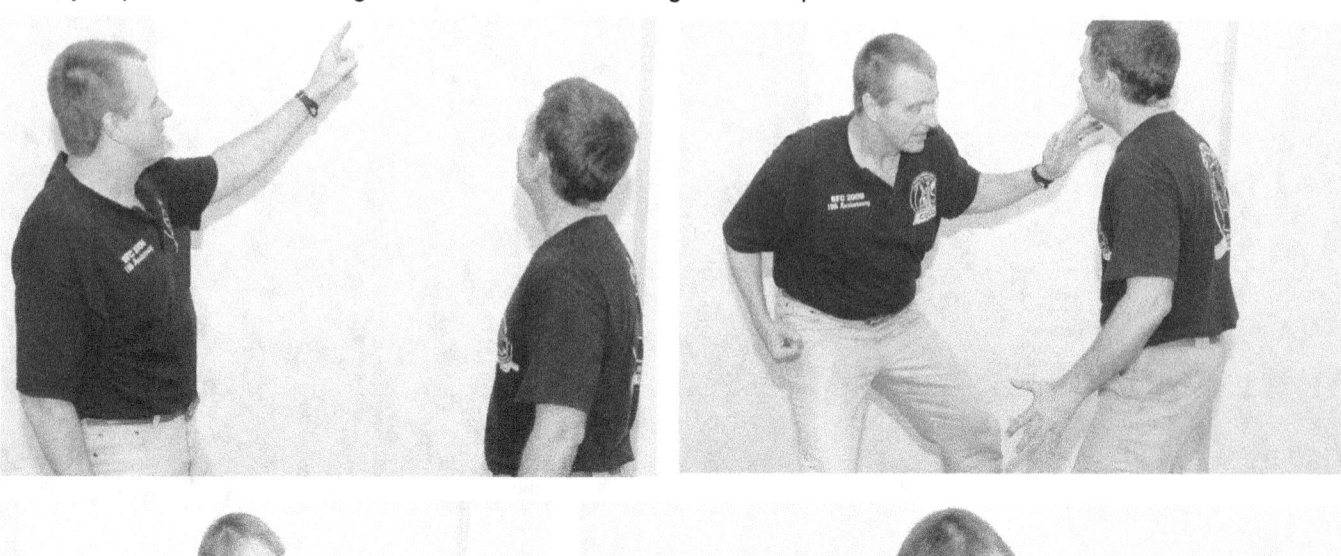

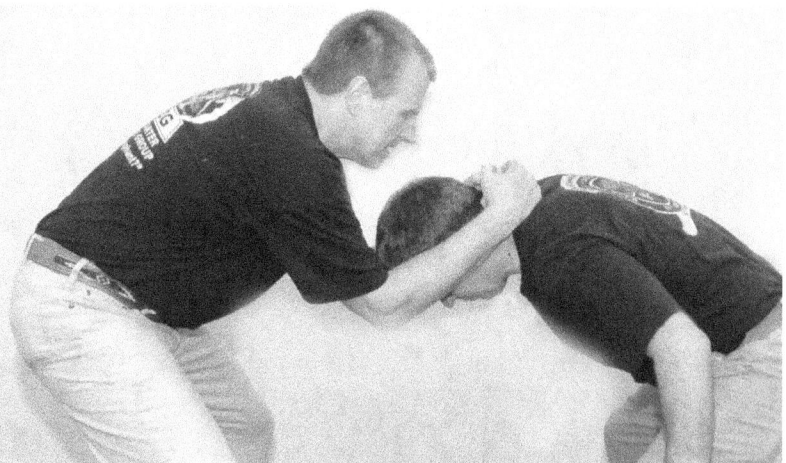

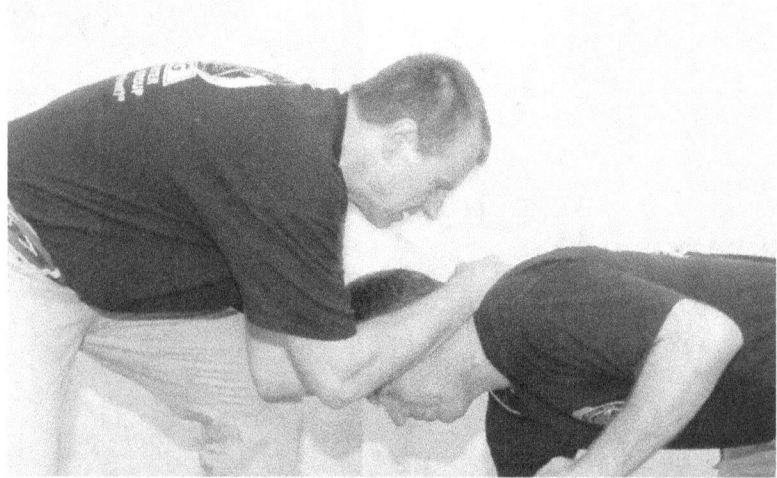

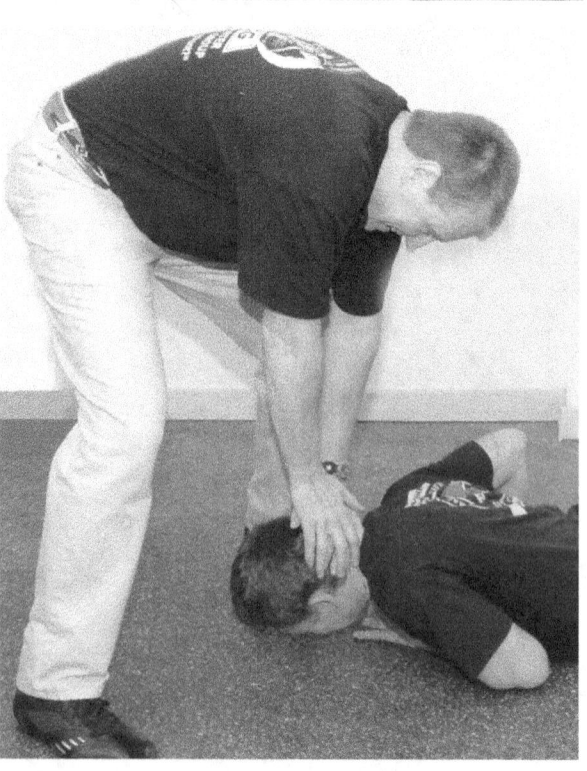

If he is stooped over and stunned, get in front of him. Knee the face if you wish or think you need to. Pull the head face down. Step back as you go, freeing the space for him to fall forward. Once he is down, take the appropriate action. Some experts suggest that you take the head down in a serpentine type of "s" pattern.

Frontal Takedowns: The Big "Splay-Down" Counter to a Tackle
One of the common street fight takedowns is a tackle. Most of the population do not know how to tackle with sport-like, practiced efficiency and tackle wildly. One of the five main counters to a tackle is this drop scenario.

He dives...

The opponent hunches over to get low and under the height of your arms. You may get your arms on his shoulder tops and slip your legs back to inhibit his ability to hook and grasp the back of your legs properly.

This leg slip is often called "splaying " the legs and body by sports specialists.

You put up your hands and get your legs back.

Then you drop as much of your torso as possible on his back. Kick your legs out even further and try to crash down onto him and flatten him hard on the ground.

If done properly, this is a devastating crash. From this point, you take the appropriate action for the situation.

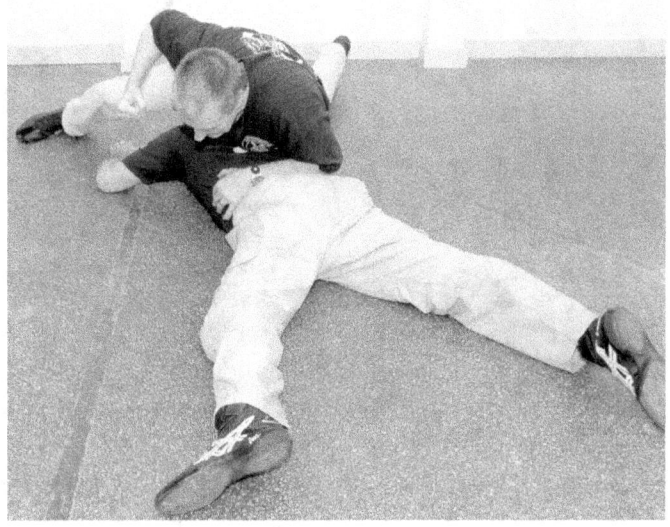

Page 75 - W. Hock Hochheim's Training Mission Five

CQCG

IMPACT WEAPONS
SDMS LEVEL 5: Impact Weapon vs. Impact Weapon Disarming and Countering / Retention

A Mission Overview

This level specifically deals with basic, "stick-versus-stick" disarming principles and the main counters to the main disarms. It would be a daunting task of thousands of pages and even more photos to detail every possible disarm and every single counter to each.

Level 9 of these CQC courses involves unarmed combatives versus weapons as an across-the-course theme so watch for it in upcoming books. This section will cover basic stick-versus-stick disarming and samples of early phase, mid-phase and late phase counters to the disarms.

There are many subjects to review in this endeavor. First, there is a difference between SMS, single-hand, single-stick opponents, and DMS-those holding their weapon with two hands. SMS and DMS disarming tactics may be somewhat different at times.

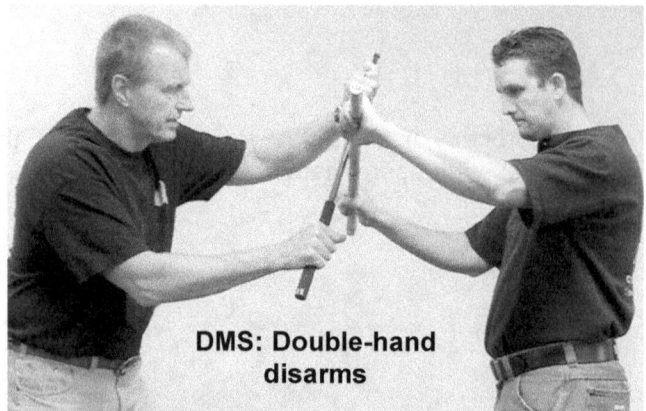

There are SMS-single hand grip disarms and DMS-double hand grip disarms.

Whether SMS or DMS, you will disarm the enemy:

1) From his weapon carry site.
2) As he draws the weapon from the carry site.
3) As he presents the weapon and threatens.
4) As he attacks with the weapon.

You must learn how to disarm in a continuum from the undrawn weapon carry stick on up to being assaulted with a fully drawn weapon..

Whether SMS or DMS, you should try and stun the enemy to complete the disarm:

1) Throw objects
2) Hand Strike
3) Kick
4) Stick (or any weapon) strike

It is imperative that your try to confuse, stun and strike the opponent. If the stunning methods seem successful? Keep stunning until they are unconscious. Disarm the weapon from an unconscious man.

It may seem useless for some to practice stick-versus-stick disarming, yet the possibilities for the need of such knowledge does occur. The stick could be an impact weapon of any type, like a flashlight or a chair leg.

Sticks, flashlights, tubes, pipes, many everyday items resemble the common stick..

European security agent Steffan Mattsson was once attacked by the detachable upper half of a medical IV carrier.

SMS: The Five Disarm Groups

It is unlikely most of us reading this book will actually be in a classic stick-versus-stick fight. We live in a mixed weapon world where your impact weapon will more likely be used versus a knife, a chair, or even an unarmed person. Still, there are many martial students and practitioners interested in the science of classical stick fighting, and part of this comprehensive study is stick-versus-stick disarming. It is something I too wanted to understand more as a basis for future variations and development.

Through some 30 years of trying to grasp and hone my stick-versus-stick disarming theories and skills, I have been through various martial, military and police approaches. None of these systems were definitive or comprehensive. Worse, most Filipino systems were not easy to grasp and poorly organized. They were almost designed to be over-complicated by their very nature. This research and experimentation has lead me to the following conclusions in single-hand, single-stick disarming. Some relate to the DMS - two handed grip disarming.

The 5 Basic SMS Impact Weapon Disarms:
   Disarm 1) Impact Disarms
   Disarm 2) Hand Snake Disarms
   Disarm 3) Stick Snake Disarms
   Disarm 4) Strip and Keep Disarms because you caught the stick.
   Disarm 5) Strip and Send Disarms because you caught the limb.

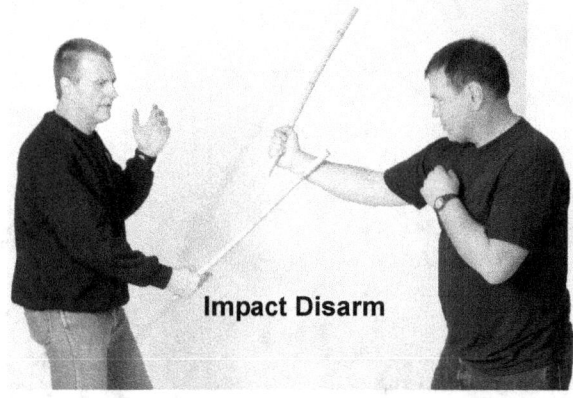
Impact Disarm

Hand Snake

Stick Snake

Strip and Keep

Strip and Send

I was challenged by an old Filipino instructor once on this 5 disarm outline. Refusing to believe that his complicated subject could be so simply defined, he grabbed a partner and did obscure disarms to prove that not all disarms fit in the 5 categories. But, as he did each one, I easily classified them. "That's a strip and keep because you caught the stick," or "that's a strip and send." And so it went until he ran out disarms.

Page 79 - W. Hock Hochheim's Training Mission Five

"On the Clock" Training Helps Pick a Side

Have your partner attack you on the four major clock quadrant angles - 12 o'clock, 3 o'clock, 6 o'clock and 9 o'clock. Let him use the shaft, then the pommel, and then the stabbing tip in sets. This is an extremely comprehensive, yet simple attack system to train against. Assign yourself a specific disarm and work them.

Generally speaking if you can do these disarms on the 3 and 9 o'clock sides, you are doing well. In the late 1980's, JKD and Filipino Martial Arts instructor Terry Gibson told me this, and it still holds true through the test of time. Terry advised me you can usually deflect or pull down an attack from above to either side of you. Do the disarm there. You can also usually deflect or pull up attacks from below to either side of you. Do the disarm there. "First learn the sides," he told me. "Then learn pulling down high attacks and pulling up low attacks."

I put this advice into practice with this clock training format I developed. There are, of course, some disarms that will work from high and low contacts. When you work through this clock assignment, you will learn them.

Strike:
1) On the four clock corners.
2) With the shaft.
3) With the handle.
4) With the pommel.

Block:
1) Unsupported block.
2) Supported block.
3) DMS block.
4) Forearm to forearm block.
5) Pass/deflect.

Practice:
1) The Impact Disarm
2) The Hand Snake Disarm
3) The Stick Snake Disarm
4) The Strip and Keep Disarm
5) The Strip and Send Disarm
6) On all four clock corners.
7) Righty vs. lefty.
8) Lefty vs. righty.
9) Experiment with and versus DMS grips.

Jeff "Rawhide" Laun attacks me from the high 12 o'clock direction.

I force the high shot to my left, or 9 o'clock side.

I force the high shot to my right, or 3 o'clock side.

He attacks low.

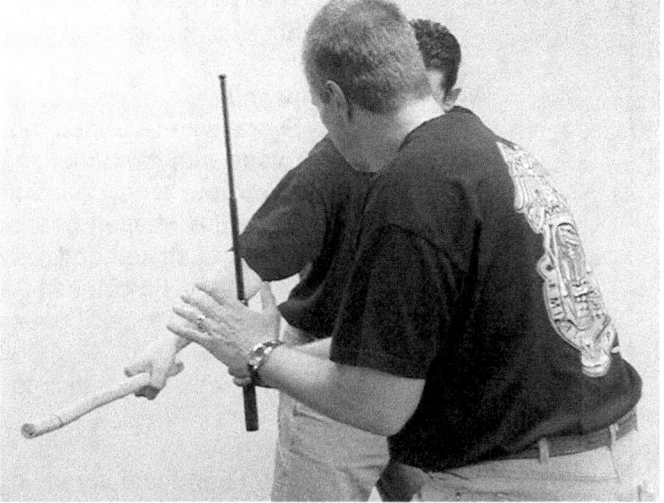

Learn to pull, or draw, his stick up to the right or left sides, comfortable positions for disarming. Obviously some of the basic disarms, such as the strip and send, can be done at 12 o'clock high or 6 o'clock low, but most disarming naturally picks a side to work best.

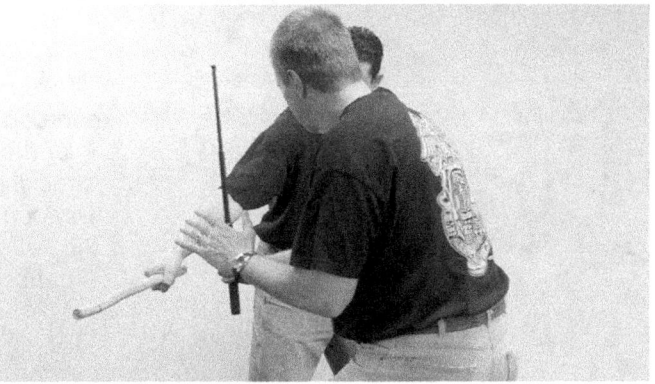

Impact Disarms: SMS Versus SMS Version

These disarms are the simplest to understand and perhaps the very best. The enemy takes a swing at you, or makes the mistake of leaving his weapon-bearing limb up and in your striking range You are in a position to strike him so hard that he loses his stick. You could indeed hit his stick so hard that he loses his grip. Or, you could strike his hand, forearm or upper arm so hard he drops the stick.

You could simply hit his body anywhere from head to toe, with such great force he loses his stick.

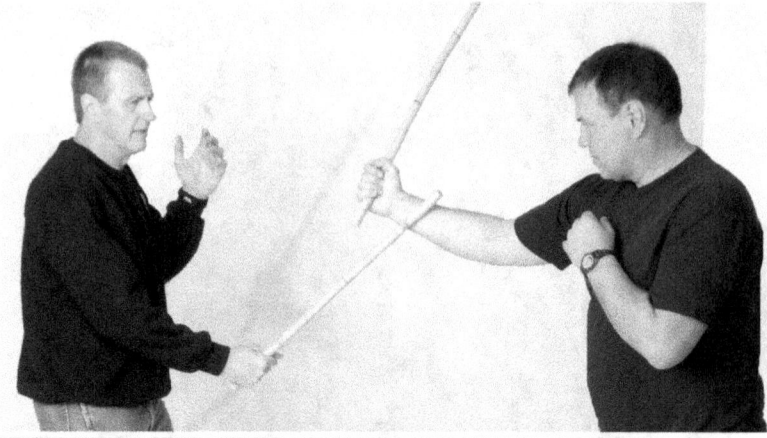

SMS Impact Weapon Disarm Practice Drills:

The trainer: Stands in threat-presence fighting stance.
Strikes on the 4 clock angles with shaft strikes.
Strikes on the 4 clock angles with tip stabs.
Strikes on the 4 clock angles with handle strikes.
Tries to draw his weapon from a carry site.
Stick vs. knife: He strikes with a knife.

With an SMS grip, you:
Evade with balanced footwork as needed.
Evade with torso motion as needed.
Keep your empty hand free of the attacking swing.
Strike the weapon bearing limb.
- strike hand, or wrist, or forearm, or best target.
- as it comes in at you.
- or, after it passes.

SMS Stick-Slide: stick-to-stick contact and slide stick down atop his hand.

Work these drills
Right hand grip versus right hand grip.
Left hand versus right hand grip.
Left hand versus left hand grip.

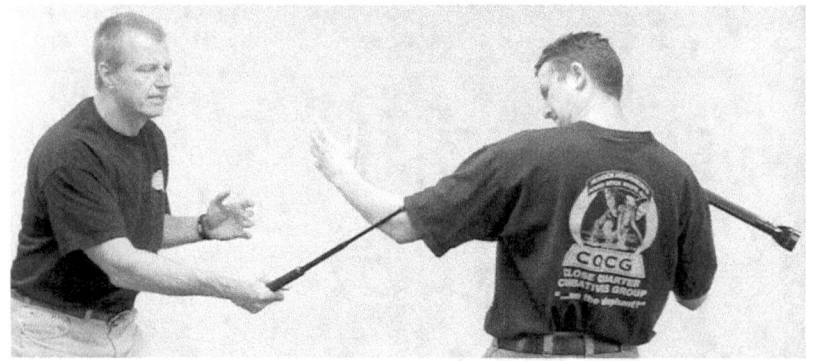

Take note that there are incidents recorded where an impact on any part of the body may be so painfully shocking that the enemy drops their weapon. An example shown here to the left. Hitting the "other" elbow has caused many a weapon disarm.

Impact Disarms: DMS Version - DMS Versus SMS and DMS

These impact disarms are the simplest to understand. The enemy takes a strike at you. You are in a position to strike him so hard that he loses his stick. You could indeed hit his stick so hard that he loses his grip. Or, you could strike his hand, forearm or upper arm so hard he drops the stick. You could hit his body, from head to toe, with such great force he loses his stick.

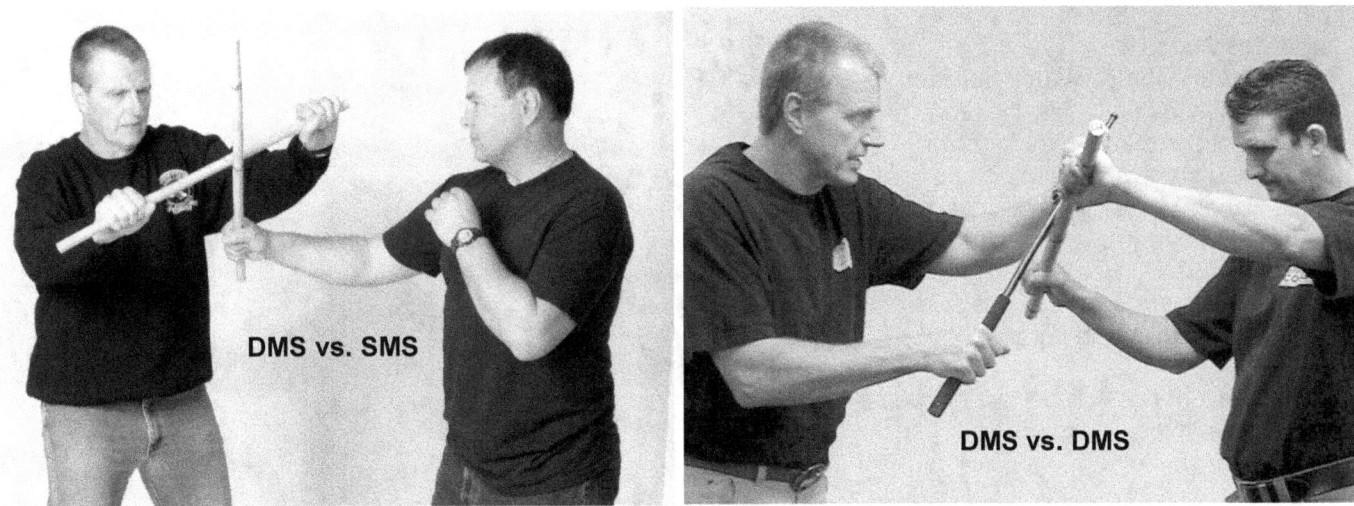

Using your DMS grip, you will be fighting against SMS and DMS.

DMS Limb Hit Practice Drills:

The trainer:
- Stands in a threatening stance.
- Strikes on the 4 clock angles with shaft strikes.
- Strikes on the 4 clock angles with tip stabs.
- Strikes on the 4 clock angles with handle strikes.
- Tries to draw his weapon from a carry site.
- Stick vs. Knife: Strikes with a knife.

With an DMS grip, each time you:
- Evade with balanced footwork as needed.
- Evade with torso motion as needed.
- Strike the weapon bearing limb.
  - strike hand, or wrist, or forearm, or best target.
  - as it comes in at you.
  - or, after it passes.
- The DMS Stick-Slide, stick-to-stick contact and slide stick down atop hand.

Work these drills:
- Versus a two-handed grip.
- Versus a person with an SMS right hand grip.
- Versus a person with an SMS left hand grip.

Impact Disarms: DMS Version versus a DMS and SMS Attack samples

Hit the Hand Directly:
You directly hit the hand, whether he has an SMS or DMS grip.

The DMS Slide:
You've made stick-to-stick contact. Your stick slides down his stick for a hand impact. You might try for both hands. You start with a DMS strike or block. Either way, stick-to-stick contact is made. You have two impact choices:

1) You can **slide** your stick against his hand to make an impact.

2) You might possibly be fast enough to slide into both hands.

Contact, then slide...

...to the hand.

If you are fast enough, slide to hit two hands.

SMS Hand Snake Disarms

The enemy takes a strike at you. You are in a position to block his stick. You wrap your free hand around his weapon bearing limb. Keep snaking until his stick is free from his hand. You may have to strike him in the process and use tricks like body positioning to assist the hand snake.

Through the years many untrained victims and even untrained criminals have reflexively hand snaked weapons from their opponents. It is virtually a reflexive movement with accidental circular follow-up motions that cause a disarm. In this stick versus stick arena, your stick contacts his stick. Your empty hand wraps the weapon side limb. The hand snakes and burrows in a clockwise or counter-clockwise motion.

SMS Practice Drills:

The trainer:
- Stands in a threatening stance.
- Strikes on the 4 clock angles with shaft strikes.
- Strikes on the 4 clock angles with tip stabs.
- Strikes on the 4 clock angles with handle strikes.
- Tries to draw his weapon from a carry site.

With an SMS grip, each time you:
- Evade with balanced footwork as needed.
- Evade with torso motion as needed.
- Keep your empty hand free of the attacking swing.
- Block his strike.
- Snake his weapon bearing limb.

Work these drills:
- Versus a person with an SMS right hand grip.
- Versus a person with an SMS left hand grip.
- Versus the 4 clock attack angles.
- Versus a two-handed grip-you hand snake one hand, strike the other.
- Versus a two-handed grip-you strike one hand, hand snake the other.

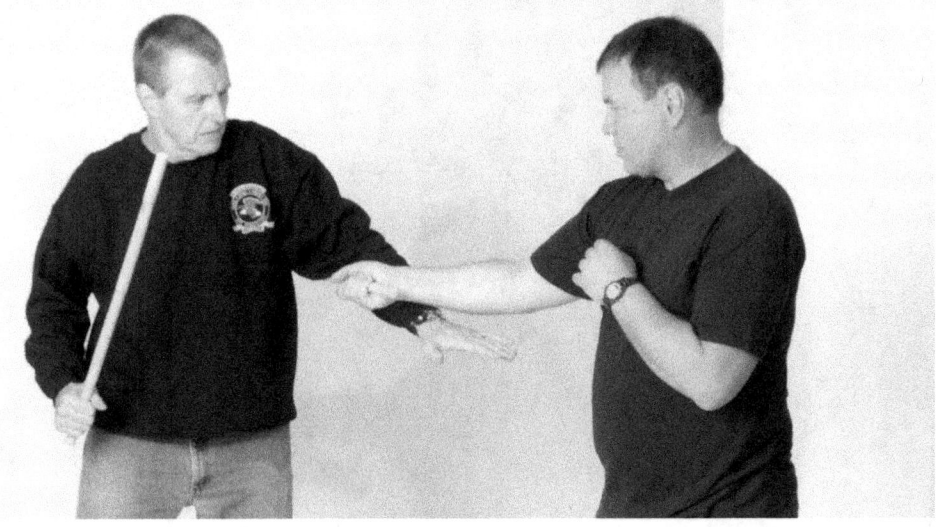

The wrap/snake might:

- catch the stick only,
- catch the connection of the hand and stick, or
- wrap the weapon bearing limb, in which case a snake disarm might not be the best tactic, and other grappling options should be pursued.

Technically if you are on the outside of the attacking arm, clockwise drilling is better. If you are on the inside of the arm, counter-clockwise drilling is better there. You may snake the shaft of the stick, above his hand. You may also snake the handle part below his hand. It is smart to strike the opponent during this process.

Hand Snake Disarm Sample: The Right-side, Clock-wise Snake

Your 3 o'clock disarm versus a backhanded attack. The clockwise snake.

Hand Snake Disarm Sample: The Left-side, Counter-clockwise Snake

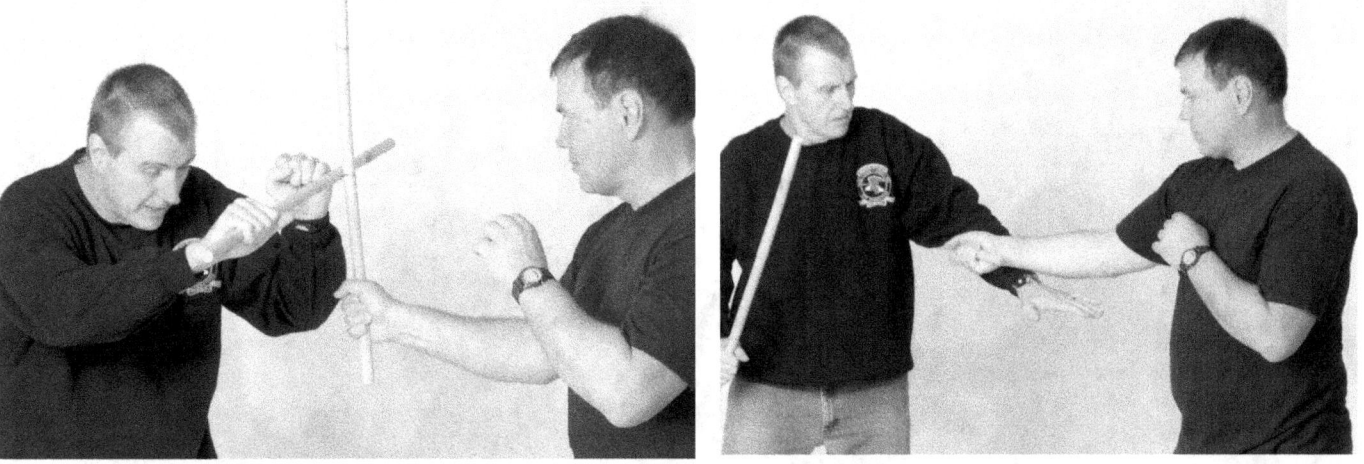

Your 9 o'clock disarm versus an inward attack. The counter-clockwise snake.

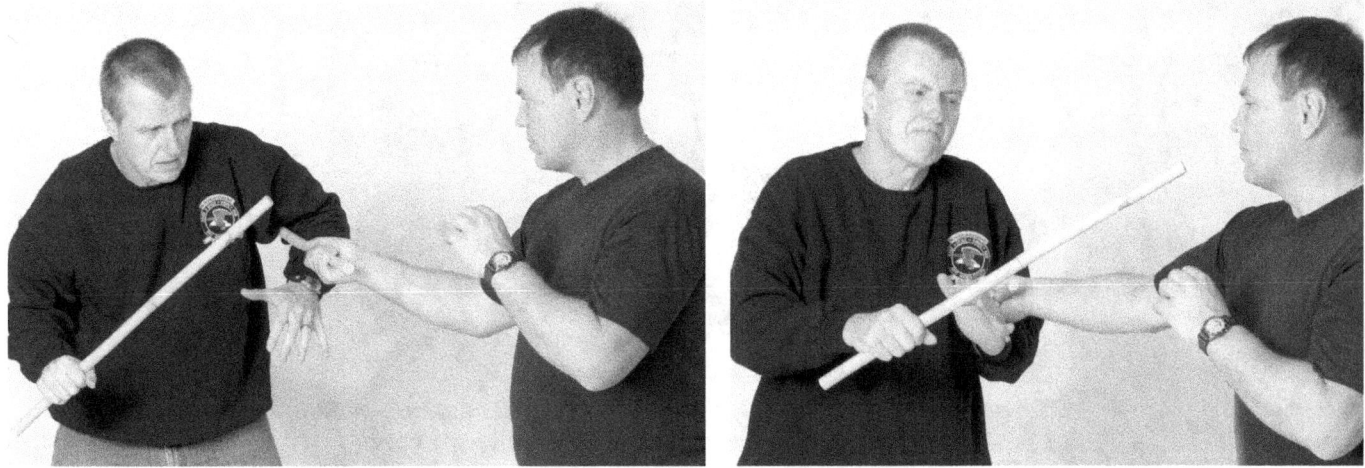

His stick is wrapped. You strike the body as you snake and pull his stick free.

Hand Snake Disarm Sample: Snake the Pommel

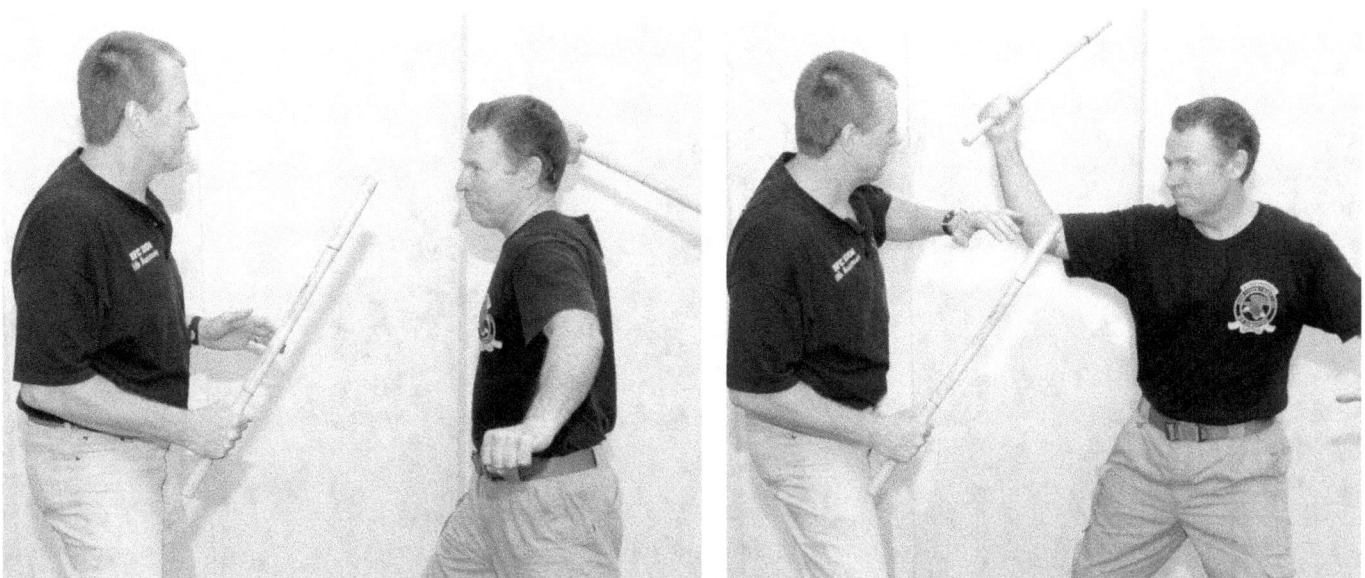

Doc Shelton attacks in close range. The viable tool is his stick pommel...

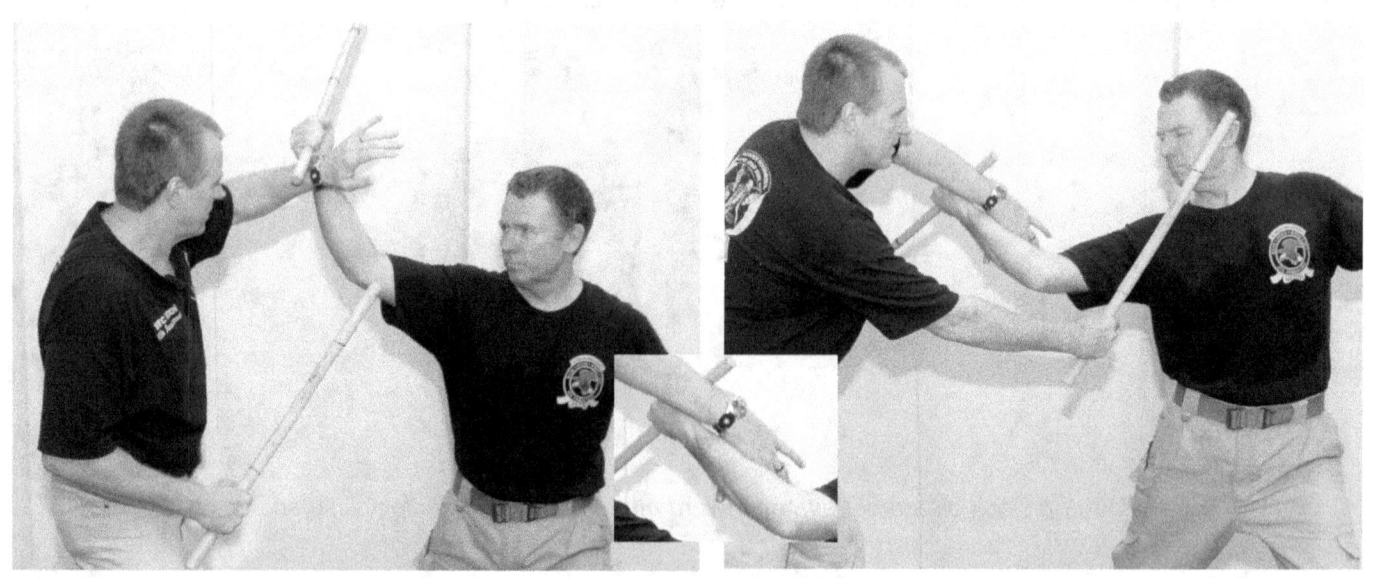

Raise your forearm high enough to engage the pommel, and start snaking. Strike him multiple times.

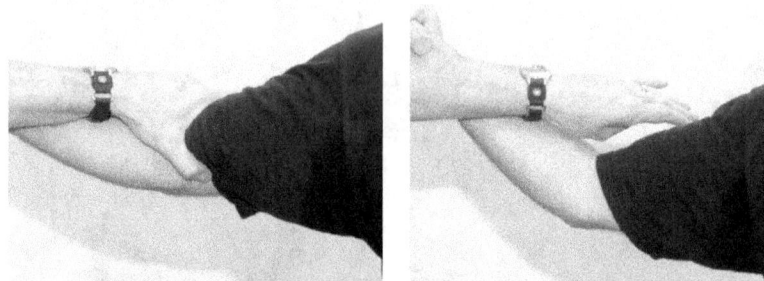

Do not hook the arm with your thumb. This will interfere with the snaking, winding invasion. Instead, tuck your thumb.

Continue to snake until the stick is released.

Hand Snake Disarm Sample: Snake the Stabbing Shaft

Doc Shelton attacks with a stick stab.

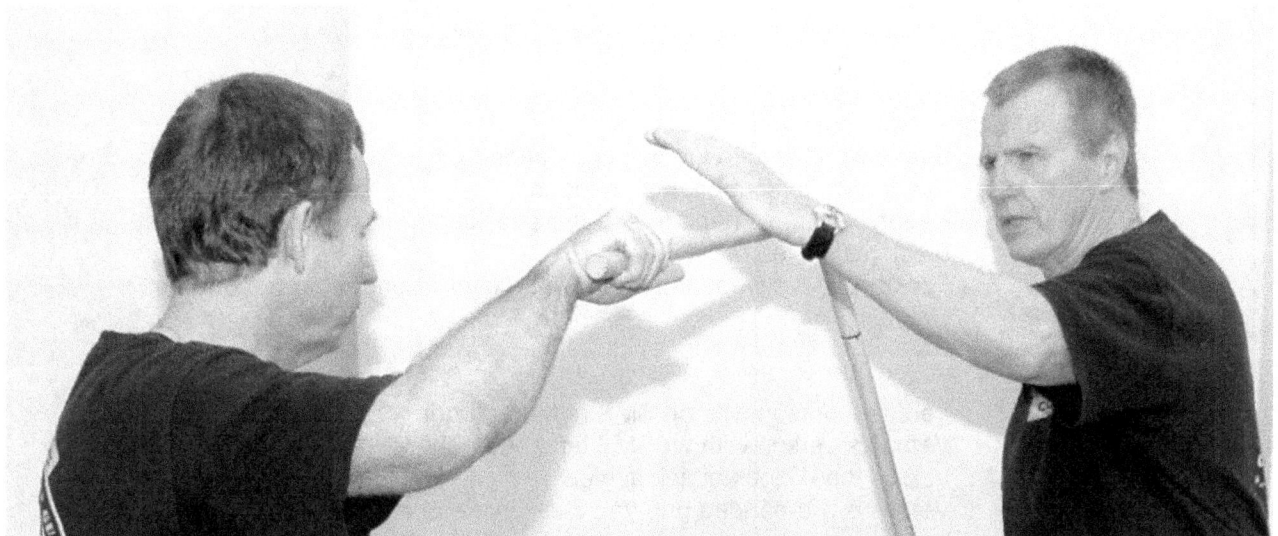

I deflect, then snake the incoming shaft of the stick, burrowing in until it comes free.

SMS Stick Snake Disarms

The enemy takes a strike at you. You are in a position to block his stick. You wrap your stick-side hand around his weapon bearing limb. Keep stick-snaking until his stick is free from his hand. You may have to strike him in the process and use tricks like body positioning to assist the stick snake.

Through the years many untrained victims and even untrained criminals have reflexively stick snaked weapons from their opponents. It is virtually a reflexive movement with accidental circular follow-up motions that cause a disarm. In this stick versus stick arena, your stick stopped his stick. Your stick-side hand/arm/stick wraps the weapon. The stick side snakes and burrows in a clockwise or counter-clockwise motion until a disarm has been completed. The following pages will display some right and left side samples.

SMS Stick Snake Practice Drills:

The trainer: Stands in a threatening stance.
Strikes on the 4 clock angles with shaft strikes.
Strikes on the 4 clock angles with tip stabs.
Strikes on the 4 clock angles with handle strikes.
Tries to draw his weapon from a carry site.

With an SMS grip, each time you:
Evade with balanced footwork as needed.
Evade with torso motion as needed.
Keep your empty hand free of the attacking swing.
Block his strike.
Stick snake his weapon bearing limb, over or under his weapon arm.

Work these drills:
Versus a person with an SMS right hand grip.
Versus a person with an SMS left hand grip.
Versus the 4 clock attack angles.
Versus a two-handed grip-you stick snake one hand, strike the other.
Versus a two-handed grip-you strike one hand, stick snake the other.

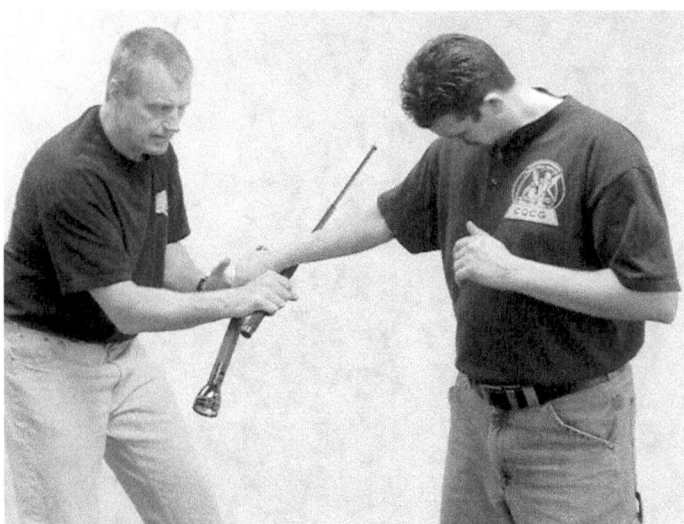

The stick snake might:

- catch the stick only.
- catch the connection of the hand and stick.
- wrap the weapon bearing limb.

Page 90 - W. Hock Hochheim's Training Mission Five

Left-Side Stick Snake Sample: The Big Arm Snake

This example is of a big sweeping arm around an inward strike. This snake captures the enemy's stick. After the block, you circle his stick arm from above. His stick is pinned across your chest.

You keep snaking the arm and twisting your torso until the stick comes free. You may or may not need to grab the weapon with your free hand.

Catch the sweet spot where his hand holds the stick. Note two diminishing shots.
One slash across the face, the other a fan on the arm.

In this big arm, stick-side snake, note how you must concentrate on capturing right where his hand and stick come together. Hit him, when possible, with your free hand or stick.

1) Block.
2) Hit back.
3) Stick arm snakes over the top. Slash face on the way.
4) Stick continues to envelop. Fan strike captured arm.
5) Capture key point where his hand holds a stick.
6) Pivot torso.
7) Continue powerful snaking and torso pivot.
8) You may or may not need to grab his stick.

Left-Side Stick Snake Sample: The Classic Vine

You stop the strike. Grab the stopped stick. Carry the tip of your stick clockwise over his wrist. This will create a bridge.

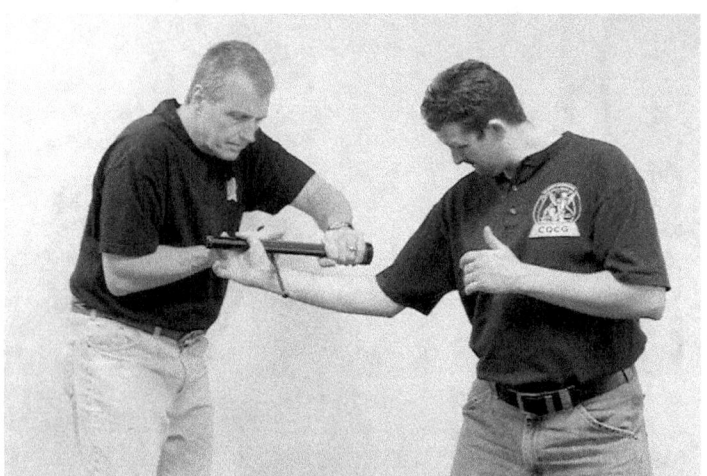

You bend his weapon over the bridge and yank the end down. You push your handle up.

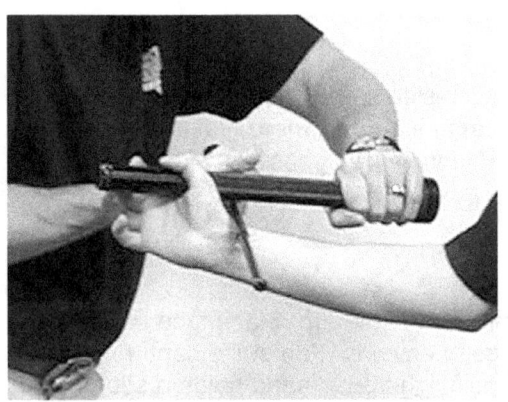

In the classic Vine Stick-Side Snake, note how you must concentrate on capturing right on top of where his hand and stick come together. Hit him when possible with your free hand or stick.

1) Block.
2) Hit back.
3) Stick snakes over the top of his forearm. Poke face.
4) Stick slips down to wrist area.
5) Capture key point atop where his hand holds a stick.
6) Push his stick down over yours.
7) Lift your hand up.

Right Side Stick Snake Sample: The Wrist Turn Over.

Stop it. Strike.

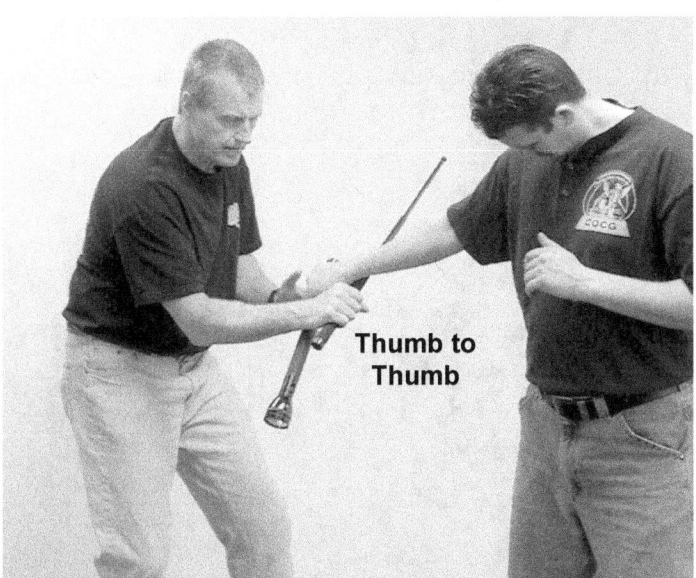

Go palms-down, thumb to thumb with him. Turn your hand and stick over palms up.

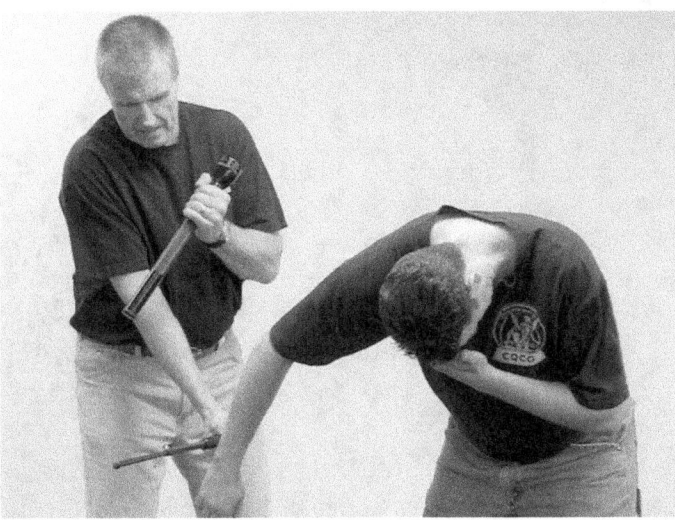

In this Wrist Turn-Over, stick snake, note how you must concentrate on getting thumb-to-thumb with him. Hit him when possible with your free hand or stick.

1) Block.
2) Hit back to the face.
3) Stick snakes under the top of his forearm.
4) Stick slips down to wrist area.
5) Go thumb-to-thumb.
6) Turn your hand over.
7) Grab and pull out his stick.

DMS Stick Snake Disarm-Statue Drill

Outside right:
  pommel snakes over wrist
  pommel snakes under wrist

Inside right:
  pommel snakes over wrist
  pommel snakes under wrist

Split:
  pommel snakes into both wrists

Inside left:
  pommel snakes over wrist
  pommel snakes under wrist

Outside left:
  pommel snakes over wrist
  pommel snakes under wrist

With a DMS 2-hand grip, there is only a stick snake. No hand is free for a hand snake. If you let loose one hand, then you can execute the prior hand snake material. Here is an educational review of the limited DMS stick snake possibilities.

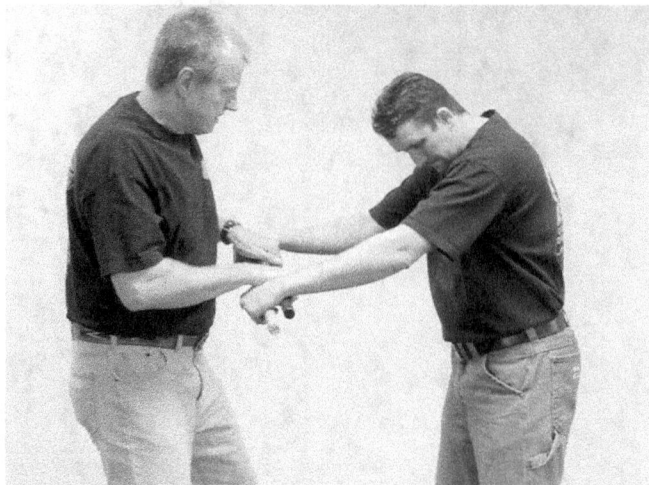

Sample: The pommel has snaked into an inside split.

Sample: You have an inside stick snake on the opponent's right side from above or below his arms.

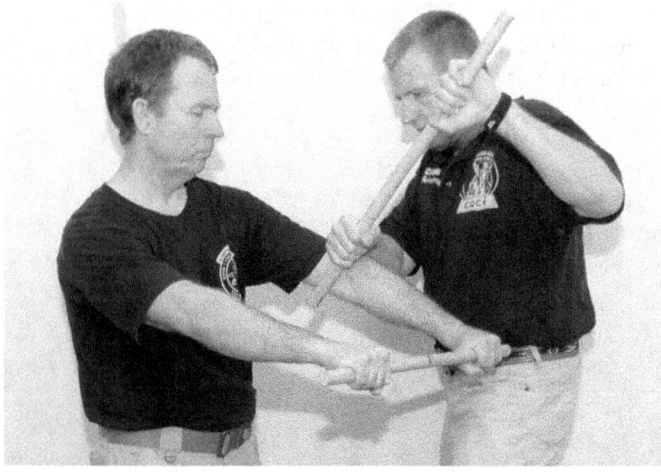

Sample: You have an inside stick snake on the opponent's left side from above or below his arms.

DMS Snake Disarm Sample: Double Snake

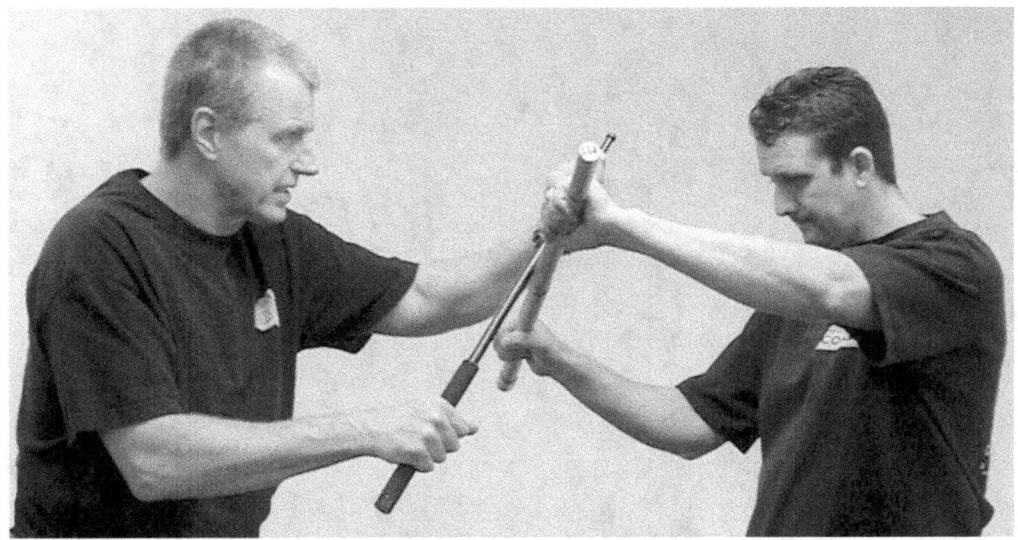

Stop or initiate DMS attack. The reference point is the same.

Attack and stun. If this looks like it is working; keep striking. Forget about the disarm.

Handle strike as you insert/snake the tip inside the arm.

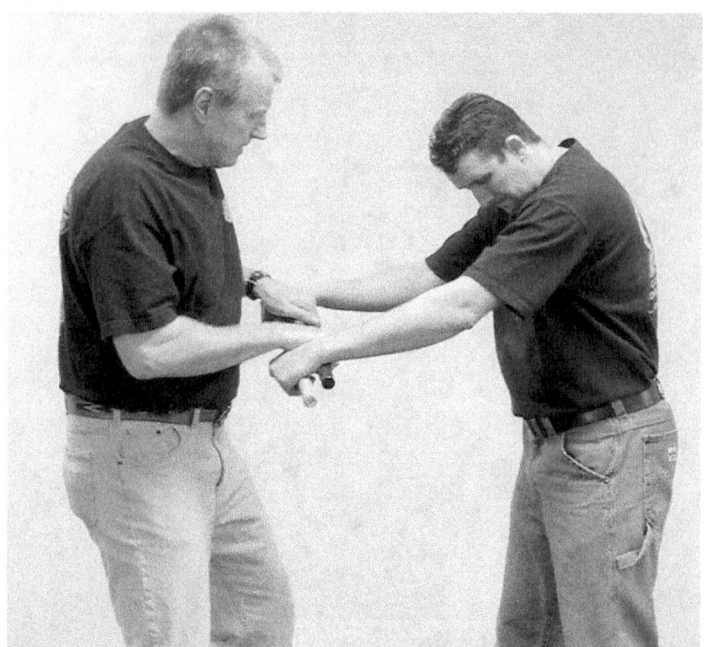

Insert the handle end inside his arm. Kick for more stunning.

Option. Some stick experts will remove their thumbs, yet capture the opponent's thumbs. This is extremely painful. This capture can be used for a throw and works quite well.

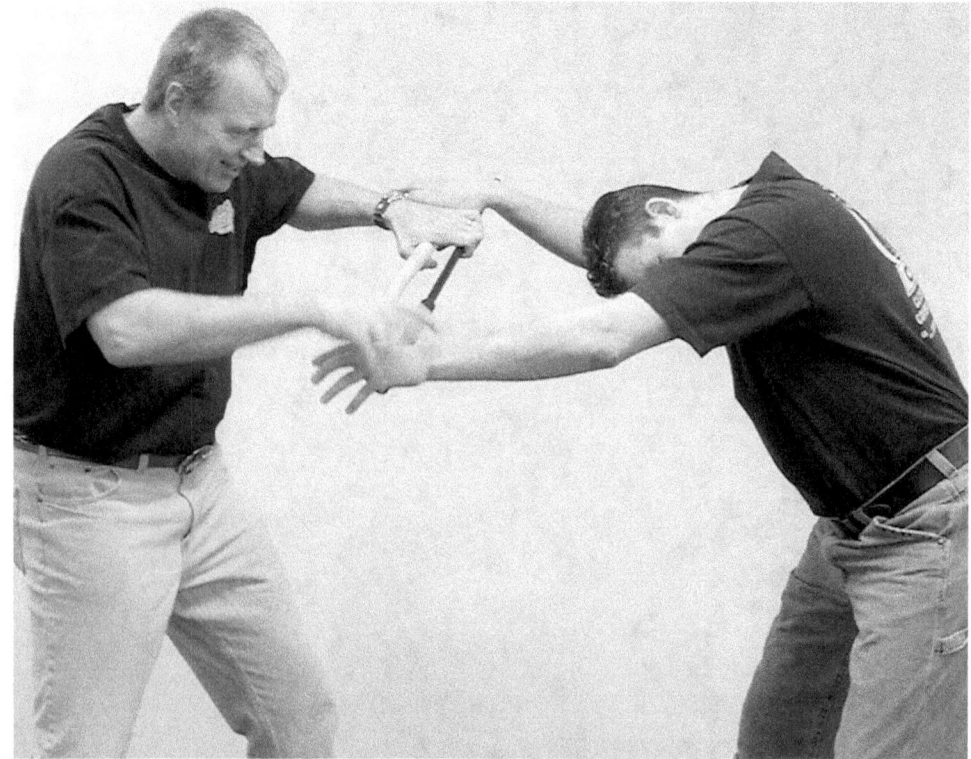
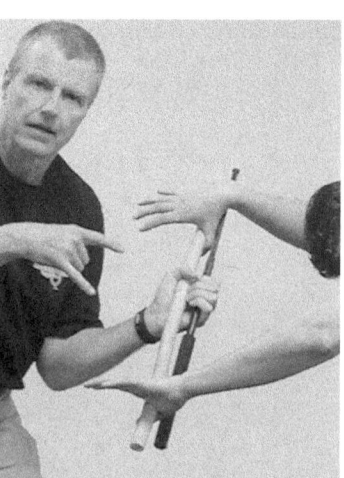

This is a painful thumb lock on both hands.

My thumbs are out of the way. His are captured.

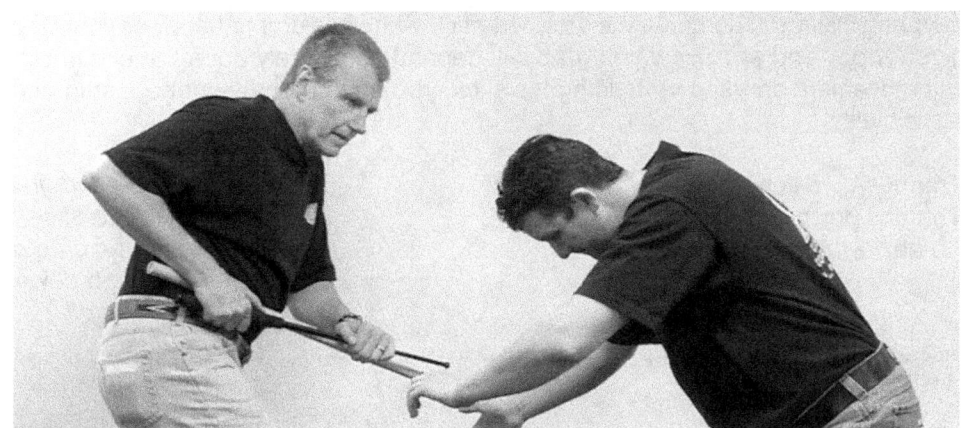

I pull out his left side handle. Use powerful body movement...

I pull out the other side, too.

Strip and Keep Disarms: SMS Versus SMS Version

The enemy takes a swing at you. You block the force. Or you strike, and he blocks the force. Either way a reference point has been established. You catch the stopped stick with your empty hand, or you may arm-wrap catch the stick between your body and arm. How you grab will depend completely upon the circumstances. Anytime you execute a push/pull movement and end up holding his stick, you have probably done a strip and keep disarm. You will seize the stick with:

Grab 1) Thumbs up catch.
Grab 2) Thumbs down catch.
Grab 3) An arm-wrap catch of the stick.

You can grab the stick (thumb up or thumb down).

Once you have captured the stick, you rip it free from the enemy's hand with:

Rip 1) *Shoot the Moon.* Hit the hand and ram your stick skyward. Pull the stick down and out the top of his hand.

Rip 2) *Slam the Earth.* Hit the hand, usually with the pommel. Ram downward and pull the stick up and out of the top of his hand.

Rip 3*) Ride the Horizon.* Rip and pull in the direction between the "moon" and "earth" mentioned above.

Rip 4) *Handle Pulls.* Hook your stick shaft or handle on his hand area and pull the stick, push the hand.

Rip 5) *Twists.* Sometimes you can twist the stick right out of his hand.

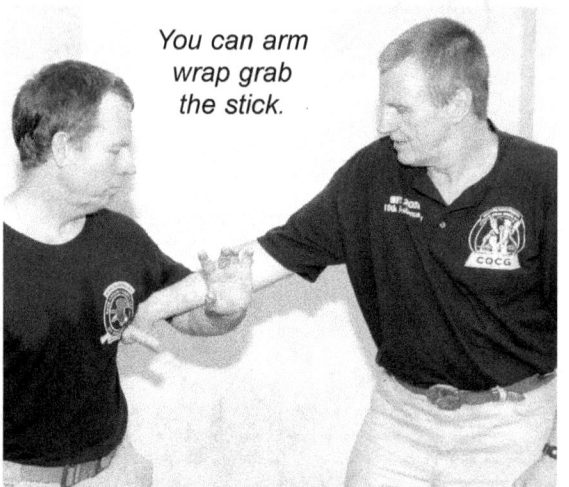

You can arm wrap grab the stick.

SMS Strip and Keep Practice Drills:

The trainer:
Strikes on the 4 clock angles with shaft strikes.
Strikes on the 4 clock angles with tip stabs.
Strikes on the 4 clock angles with handle strikes.
Stands in a fighting stance.
Tries to draw his weapon from a carry site.

With an SMS grip, each time you:
Evade with balanced footwork as needed.
Evade with torso motion as needed.
Catch his stopped stick with your hand or arm wrap.
Hit the hand/wrist/arm or even the body as you rip and/or twist the stick free.

Work these drills
Right hand grip versus right hand grip.
Left hand versus right hand grip.
Left hand versus left hand.

Strip and Keep Sample 1: Shoot the Moon

Contact made.

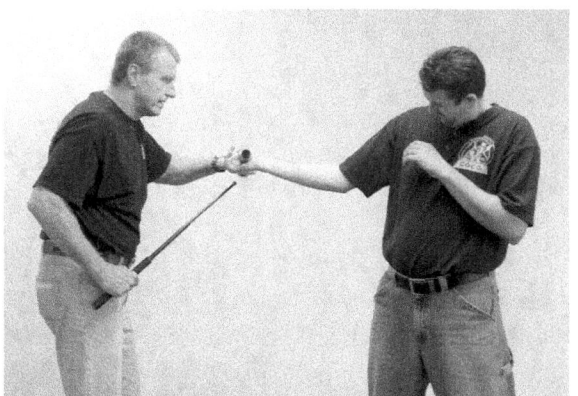
You catch the stick. You hit his limb.

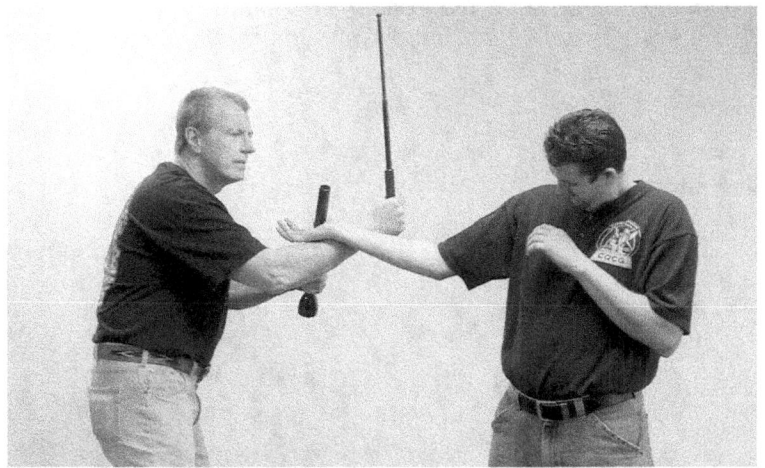
You ram your stick up and pull his stick down.

Strip and Keep Sample 2: Slam the Earth

You have stopped the stick. You catch the stick with your free hand. Your stick arm goes over the top of his arm. Your pommel strikes his upper hand or wrist. You drive his weapon limb down as you rip and/or twist his stick from his injured hand.

Strip and Keep Sample 3: Ride the Horizon

 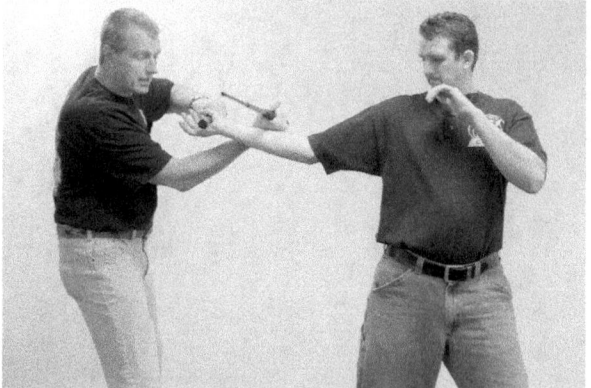

You have stopped the stick. You catch the stick with your free hand. Your stick arm goes under his arm and stabs over the top of his arm. You drive his weapon limb out as you rip and/or twist his stick from his injured hand. This rip-out travel between the "moon" and the "earth," ergo the name Ride the Horizon.

Strip and Keep Sample 4: Handle Pulls

 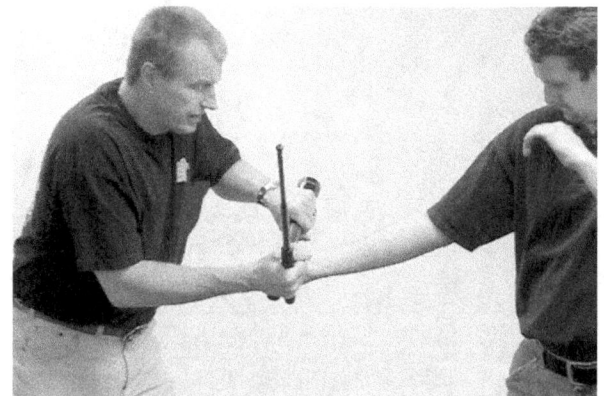

You have stopped the stick. You catch the stick with your free hand. Strike him. Your stick arm goes over the top of his arm. You hook your handle over his pommel or over his stick shaft, and pull it free.

Strip and Keep Sample 5: Twist Outs

You have stopped the stick. You catch the stick with your free hand. Strike him. Your hand starts to twist out the stick as you strike his arm.

Some Strip and Keep Disarm Training Drills

Isolated Catch Developing Drills:
    Set 1) You block and catch
        1) He strikes on all four combat clock corners.
        2) You block.
        3) You grab his stick.
        4) Practice until greater speed is achieved.

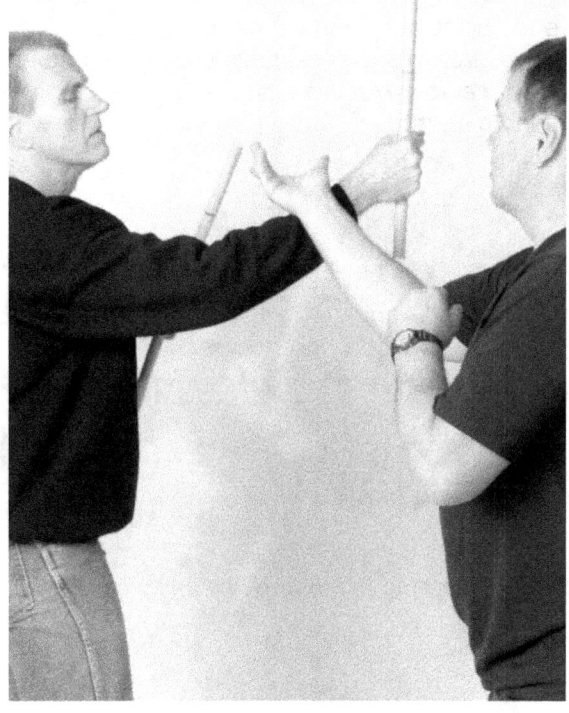

    Set 2) You strike and catch
        1) You strike on all four combat clock corners.
        2) He blocks.
        3) You grab his stick.
        4) Practice until greater speed is achieved.

Isolated Catch and Strike Developing Drills:
    Set 1) You block, catch and strike
        1) He strikes on all four combat clock corners.
        2) You block.
        3) You grab the middle of his stick and hit him with his own stick.
          (Note: not very effective on his low strikes.)
        4) Practice until greater speed is achieved.

    Set 2) You strike and catch
        1) You strike on all four combat clock corners.
        2) He blocks.
        3) You grab the middle of his stick and hit him with his own stick.
          (Note: not very effective on your low strikes.)
        4) Practice until greater speed is achieved.

Catch, Strike and Disarm Developing Drills:
    Set 1) You block, catch and strike
        1) He strikes on all four combat clock corners.
        2) You block.
        3) You grab the middle of his stick and hit him with his own stick.
          (Note: not very effective on his low strikes.)
        4) Disarm.
        5) Practice until greater speed is achieved.

    Set 2) You strike and catch
        1) You strike on all four combat clock corners.
        2) He blocks.
        3) You grab the middle of his stick and hit him with his own stick.
          (Note: not very effective on your low strikes.)
        4) Disarm.
        5) Practice until greater speed is achieved.

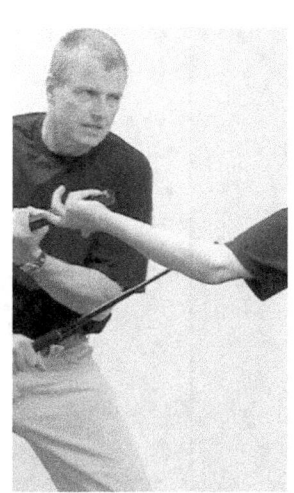

SMS Disarms: The Strip and Send Disarms:

The enemy takes a swing at you. You block the force. Or you strike, and he blocks the force. Either way a reference point has been established. You catch his empty hand/wrist/arm with your empty hand, or you may arm-wrap catch his arm between your body and arm. How you grab will depend completely upon the circumstances. Anytime you execute a push/pull movement and end up holding his arm, you have probably done a strip and send disarm.

Your grabs to catch:
   Grab 1) Thumbs up catch.
   Grab 2) Thumbs down catch.
   Grab 3) An arm-wrap catch of the arm.

Catching his thumb is a lucky catch. Pull it, push stick.

Push to send with:
   1) Your stick shaft.
   2) Your stick pommel.
   3) Your hand, forearm, upper arm.
   4) Your torso.
   5) Even your leg.
   6) You may even send the stick by harshly striking it with your stick.

Once the arm is caught, your can use parts of your body or your stick to strip and send his.

SMS Strip and Send Practice Drills:

The trainer:
   Strikes on the 4 clock angles with shaft strikes.
   Strikes on the 4 clock angles with tip stabs.
   Strikes on the 4 clock angles with handle strikes.
   Stands in a fighting stance.
   Tries to draw his weapon from a carry site.

With an SMS grip, each time you:
   Evade with balanced footwork as needed.
   Evade with torso motion as needed.
   Catch his stopped hand, wrist, forearm with your hand or arm wrap.
   Hit and push his stick free from his grip.

Work these drills:
   Right hand grip versus right hand grip.
   Left hand versus right hand grip.
   Left hand versus left hand.

Some Strip and Send Disarm Training Drills

Isolated Catch Developing Drills:
    Set 1) You block and catch.
        1) He strikes on all four combat clock corners.
        2) You block.
        3) You grab his limb.
        4) Practice until greater speed is achieved.

    Set 2) You strike and catch.
        1) You strike on all four combat clock corners.
        2) He blocks.
        3) You grab his limb.
        4) Practice until greater speed is achieved.

Isolated Catch and Strike Developing Drills:
    Set 1) You block, catch and strike.
        1) He strikes on all four combat clock corners.
        2) You block.
        3) You grab his limb and hit him.
        4) Practice until greater speed is achieved.

Catch, Strike and Disarm Developing Drills:
    Set 1) You block, catch and disarm.
        1) He strikes on all four combat clock corners.
        2) You block.
        3) You grab his limb.
        4) Disarm.
        5) Practice until greater speed is achieved.

    Set 2) You strike, catch and disarm.
        1) You strike on all four combat clock corners.
        2) He blocks.
        3) You grab his limb.
        4) Disarm.
        5) Practice until greater speed is achieved.

Weapon Retention and Counters to Stick Disarms

Practice the big five disarms on the four attack angles of the clock. To experiment with counters and retention, try the list below on each angle of each one.

Counters in Time Phases

There are:

1) Early-Phase Counters
2) Mid-Phase Counters
3) Late-Phase Counters

Common Counters are:

1) Explosive Retraction.
2) Hit the Disarmer.
3) Kick the Disarmer.
4) Hand Switch.
5) Slap Release.
6) Handle Punch.
7) Row and Roll Releases.
   Rowing sticks and rolling elbows
      a) - vs. stick grabs
      b) - vs. arm grabs
8) Charge in and blast away when stick has been lost.
9) Unique solutions to unique problems.

1) Explosive retraction. Usually best in the early phase. Reflexive.

2) Hit the disarmer.

3) Kick the disarmer.

4) Hand switch at a key moment.

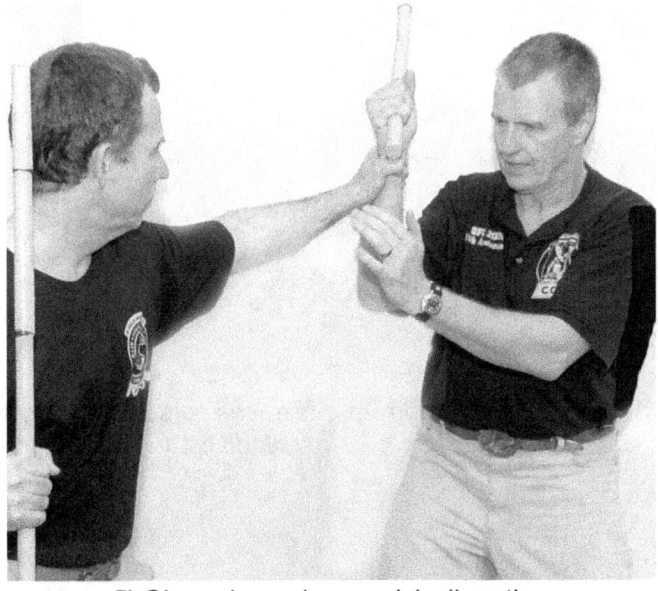
5) Slap release in a push/pull motion.

6) Handle punch your snaked stick or grabbed stick.

7a) Row and rolling releases versus an arm grab - here is a wrist roll-over.

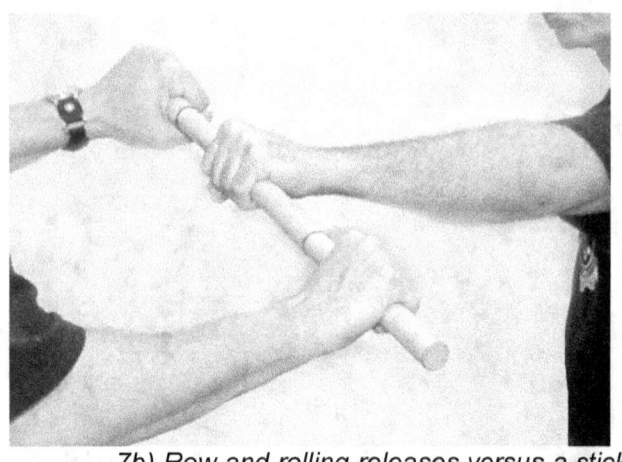 

7b) Row and rolling releases versus a stick grab. Your forearm hits his forearm. His arm has been rolled into a weakened "s," "v" or "center" lock position.

8) Loose the stick? Charge like a striking madman!

Quick Disarm Counters Study List

Impact Disarm
    Explosive Retraction.   Hit the Disarmer.   Kick the Disarmer.   Hand Switch.
    Charge in.   Unique solution - pass and hit.

Snake Disarm
    Explosive Retraction.   Hit the Disarmer.   Kick the Disarmer.   Hand Switch.
    Handle Punch.   Charge in.
    Unique solution: Grasp wrist of hand snake and throw (select a grappling technique).

Stick Snake
    Explosive Retraction.   Hit the Disarmer.   Kick the Disarmer.   Handle Punch.
    Charge in.

Strip and Keep Disarm
    Explosive Retraction.   Hit the Disarmer.   Kick the Disarmer.   Hand Switch.
    Slap Release.   Handle Punch.   Row and Roll Releases.
    Charge in.

Strip and Send Disarm
    Explosive Retraction.   Hit the Disarmer.   Kick the Disarmer.   Hand Switch.
    Slap Release.   Row and Roll Releases.   Charge in.

Keep developing this list on your own...

Basic Counters Unique to the Impact Disarm

The basic counters are:
1) Evade his impact.
2) Evade his impact and then counter-strike.
3) Switch hands, if your arm is really hurt or partial hurt.
4) Loose the stick? Charge him like a madman.

Counter 1) Evade the Impact by retracting your weapon and weapon-bearing limb away.

If you see the impact disarm coming in?

Retract your weapon and arm from his incoming strike.

Practice Drills
a) Trainee stands in a ready stance and trainer strikes at hand.
b) Trainee strikes with:
- shaft attack on all four clock angles.
- tip stab attack on all four clock angles.
- handle attack on all four clock angles.

c) Trainer tries to hit trainee's incoming strike.
d) Trainee retracts his weapon and limb away from this impact.
e) Work right hand versus left. Work left hand versus left.

Counter 2) Evade his impact and then counter-strike.

You attack, but you see his impact disarm strike coming in on you. You manage to avoid the strike.

He will do either one of two things, he will miss you and pass by your arm, as shown to the right here in the photo sequence, or he will also retract his stick. You must strike his limb either way.

He misses you and:
- passes you
- or, retracts back.

Practice Drills:

a) Trainee stands in a ready stance.

b) Trainee strikes with shaft, tip and handle on all four clock angles.
- shaft attack
- tip stab attack
- handle attack

c) Trainer tries to hit trainee's incoming strike.

d) Trainee retracts his weapon and limb away from this impact.

e) Trainer passes the intended impact point, and trainee strikes the trainer.

f) Trainer retracts his stick and limb away from the intended impact point and the trainee strikes the trainer.

g) Work right hand versus left. Work left hand versus left.

I see the incoming strike.

I avoid the impact by pulling my arm back.

I strike his missed strike. This time his momentum carried him past his intended target point.

Counter 3) Switch hands if your arm is really hurt or partially hurt.

The trainer strikes and lands. It hurts! Bad enough to drop the stick or partially drop the stick because your hand is swollen and numb, ruptured or even bleeding. You must switch hands.

Practice Drills:

a) Trainee stands in ready stance.

b) Trainee strikes with shaft, tip and handle on all four clock angles.
- shaft attack
- tip stab attack
- handle attack

c) Trainer successfully strikes your limb. You hand switch. Note-start one set with a right hand grip, one set with a left hand grip. This way at the end of each set, you have completed a right hand to left hand, and left hand to right hand cycle.

Trainee switches hands and keeps fighting.

Note: This can be practiced in a continual drill format. You strike at 12, get hit, switch hands and strike with this hand at 3. You get hit, switch hands and strike at 6. You get hit, switch hands, and strike at 9.

Next start the process with at 12 o'clock but with your left hand. This drill covers all left hand and right hand switch skills.

Retaining the Baton: Countering Your Interrupted Belt-Carry Quick Draw

Many police officers, military personnel and security officers carry batons. Many are issued "straight" simple sticks that are carried on their uniform belts in partial leather holsters, or hung by rings on straps. If the officer carries a pistol on his strong side, he will carry his baton on his weak side. Many officers and guards around the world do not carry firearms and are issued impact weapons. In this case, these non-firearm personnel will carry their sticks on their strong side. These are issues discussed in *Training Mission One.*

Frequently, the opponent will reach for an officer's impact weapon to stop him from drawing it. He will grab the stick, or he will grab your hand going for your stick. He will reach with his right hand or left hand to your right side or left side.

This right or left belt carry-site, and right or left grab-hand makes for a mix and match that creates a mathematical puzzle. The following are some explanations and sample solutions to that puzzle.

The Belt Carry Quick Draws:

Left-to-left draw   *Right-to-left draw*   *Right-to-right draw*   *Left-to-right draw*

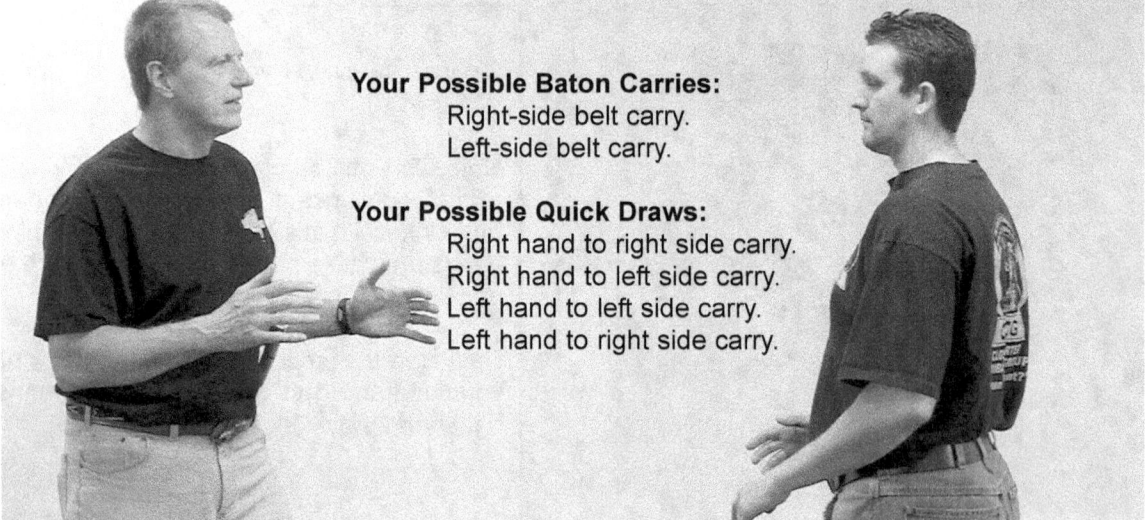

Your Possible Baton Carries:
Right-side belt carry.
Left-side belt carry.

Your Possible Quick Draws:
Right hand to right side carry.
Right hand to left side carry.
Left hand to left side carry.
Left hand to right side carry.

Sample 1: A circular wrist release from a same-side grab.

1

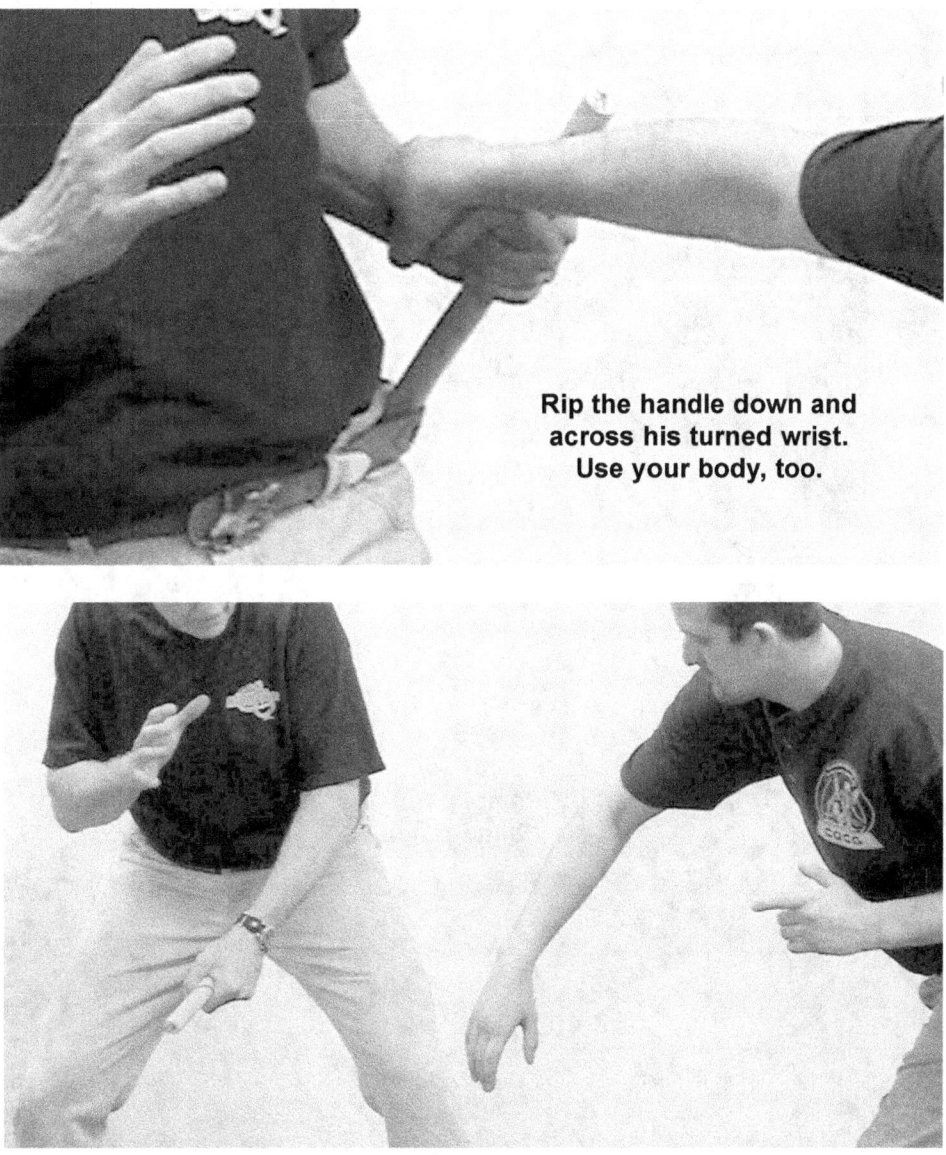

Rip the handle down and across his turned wrist. Use your body, too.

Page 111 - W. Hock Hochheim's Training Mission Five

Sample 2: A circular wrist release from a cross-side grab.

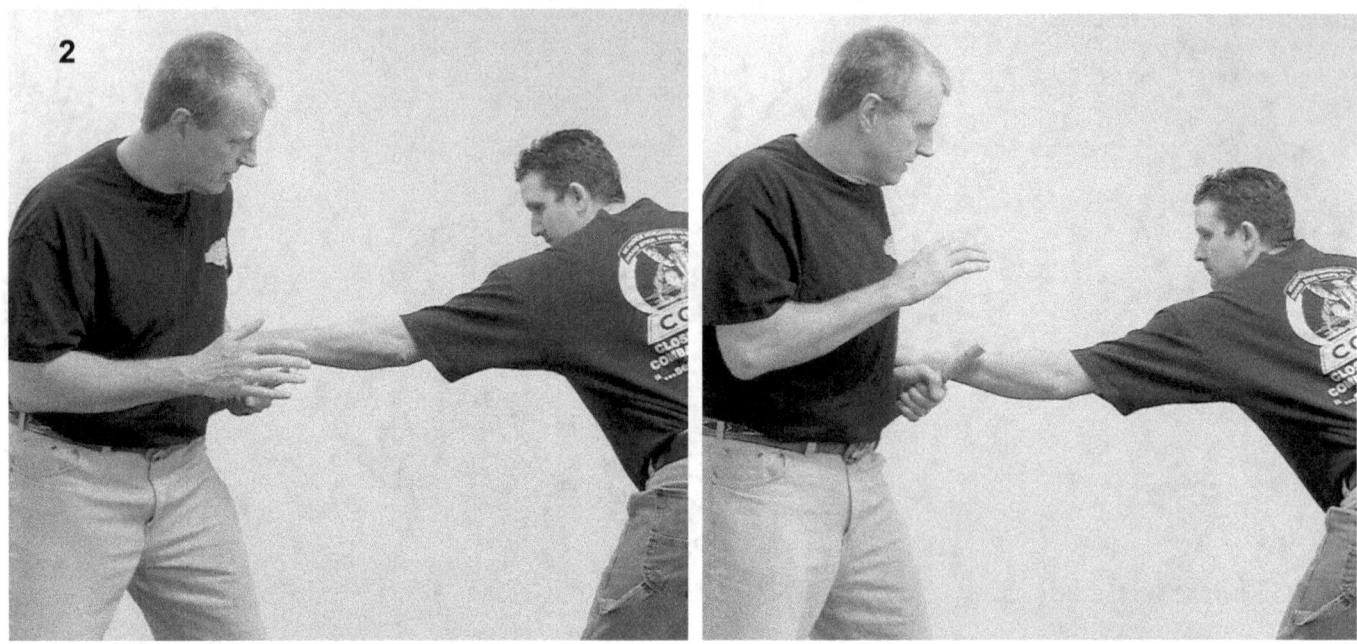

Opposite side "hand-shake" grab. My same-side pull. I start the pommel under his forearm.

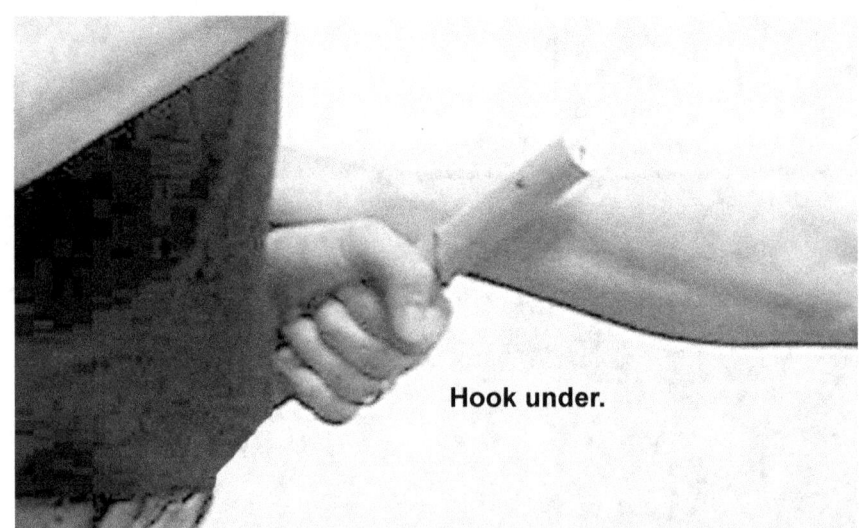

Hook under.

Pommel push and hand push.

Sample 3: An elbow roll-over release from a grab.

The classic elbow roll-over works quite well. As with many of these positions, we try to put the opponent into the weakened "s," "v," or center-lock position.

Roll your elbow over the top and then shove it down hard. Drop your body weight also if need be.

A "palm-down grab." I have a cross-draw pull. I can't get the pommel under his forearm.

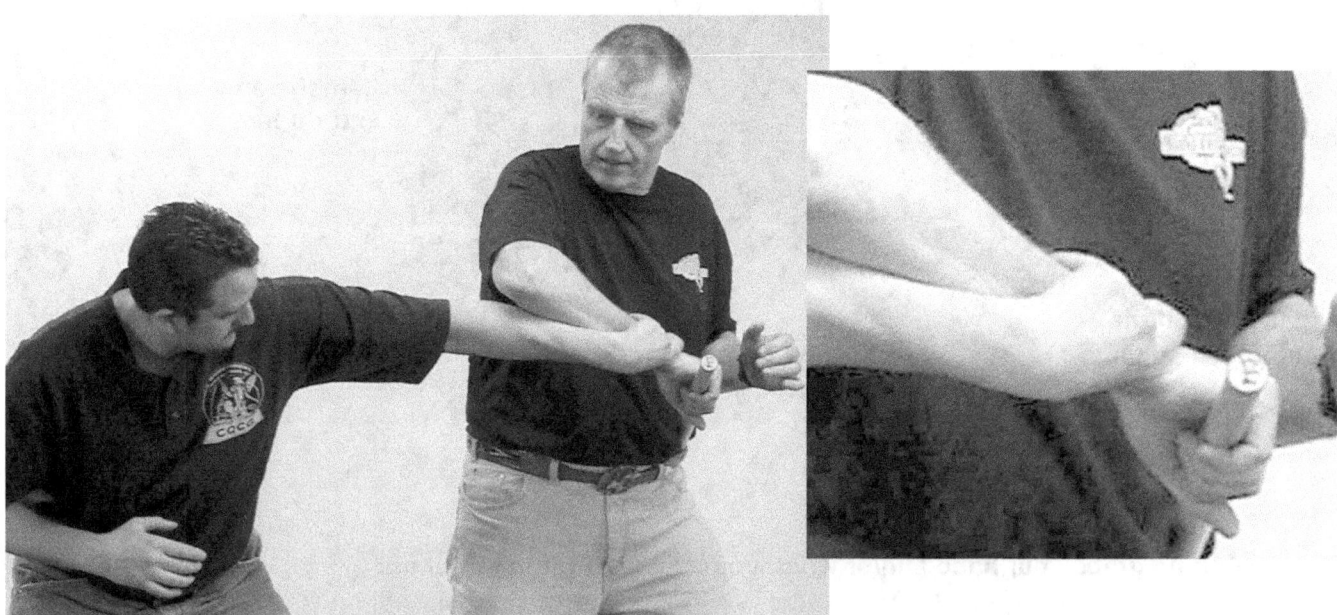

Drop down.

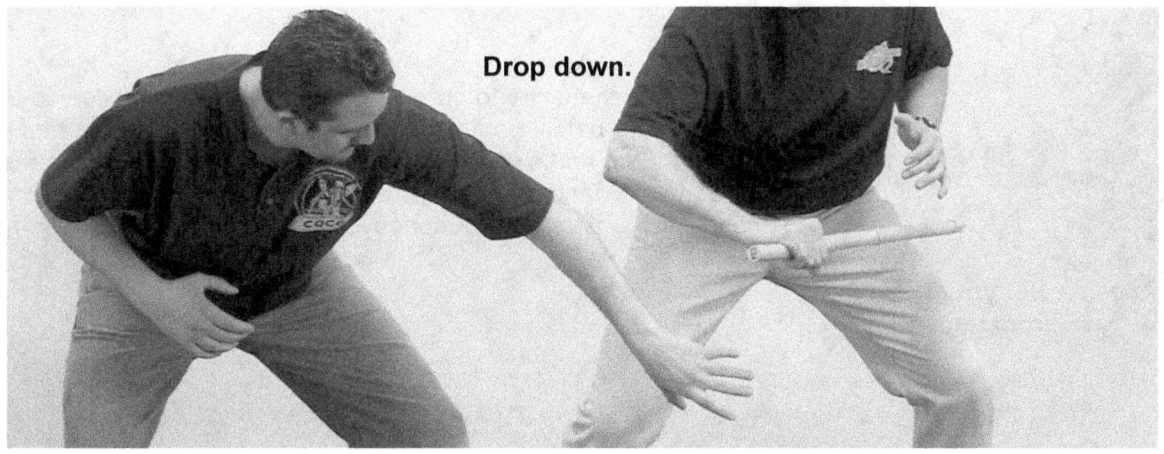

Sample 4: A wrist rip - release from a cross-side grab.

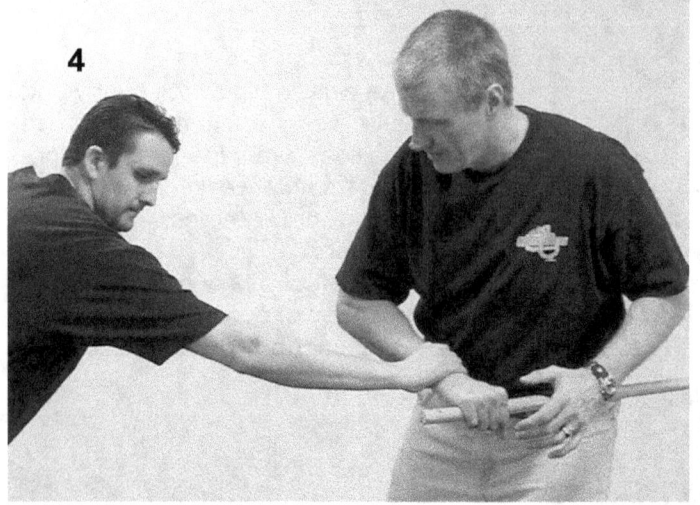

Another palm-down grab. I have a cross-draw pull. I can get the pommel under his forearm.

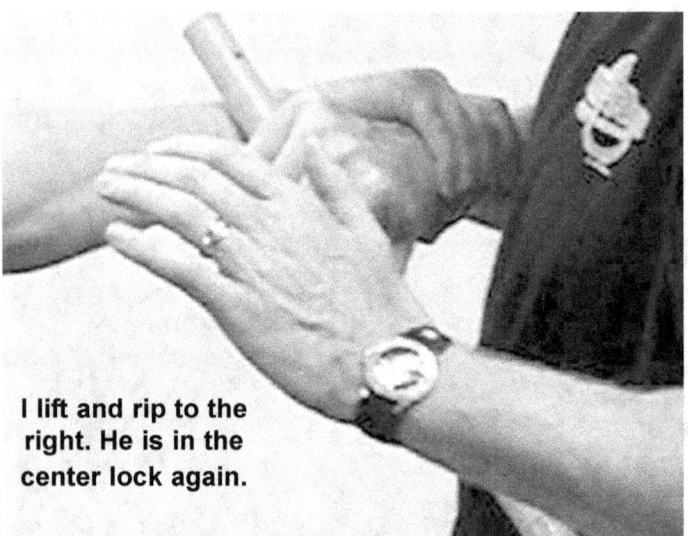

I lift and rip to the right. He is in the center lock again.

Slam the arm and rip free.

Sample 5: If he grabs your hand and/or wrist, you may lift it off the weapon.

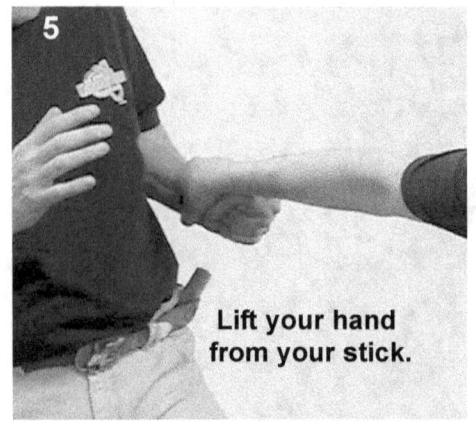

Lift your hand from your stick.

Should the enemy try to grab your hand or forearm while you are trying to draw your baton, you might let go of your baton handle and raise your hand up and away from the handle. Now this becomes a simple grab of your hand and not a wrestling match over your weapon. Your hand will be more free to maneuver to escape.

Sample 6: He grabs your same-side draw as you partially pull your baton out.

You draw. You get your stick out and almost up into action, but he grabs the shaft.

You resort to a two-handed DMS grip.

The actors switch sides to best display the action. Leery of his other hand, you step off and make an outward turn of your stick. This hyper-extends his wrist. It weakens his grip. You hit his forearm with you forearm in this process.

Remember:

* left hand carry with a left hand draw = results in a reverse stick grip presentation.
* right hand carry with a right hand draw = results in a reverse stick grip presentation.
* left hand carry with a right hand cross draw = results in a saber stick grip presentation.
* right hand carry with a left hand cross draw = results in a saber stick grip presentation.

Sample 7: A push/pull release.

A push/pull release is often successful and worthy of practice. But, this involves your two hands and may open you up to a punch. Practicing these counters with one hand is a wise choice.

Retaining the Baton:
Countering the Interrupted Belt-Carry, Quick Draw, A Summary:

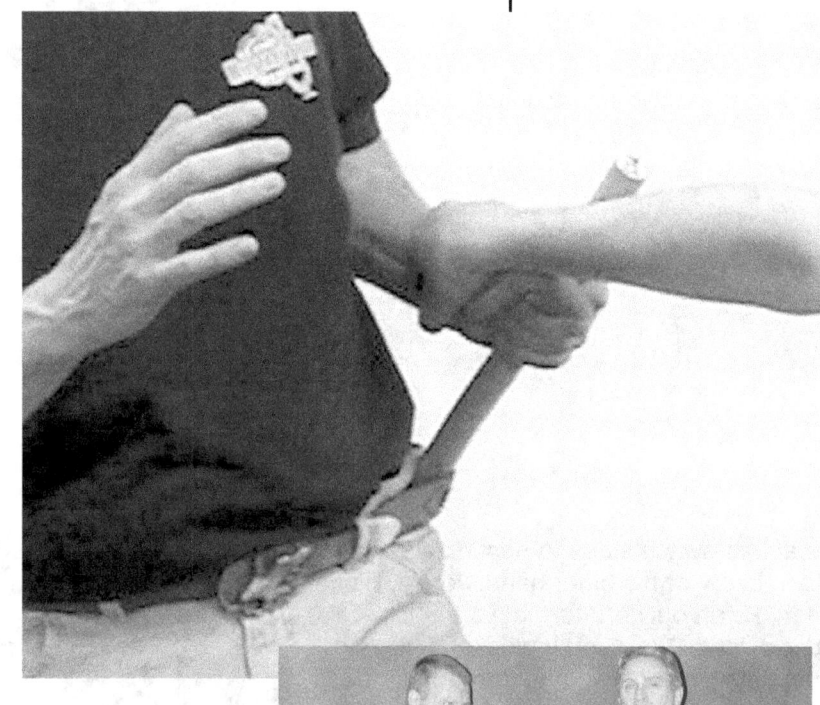

Your Possible Baton Carries
    Right-side carry.
    Left-side belt carry.

Your Possible Quick Draws
    Right hand to right-side carry.
    Right hand to left-side carry.
    Left hand to left-side carry.
    Left hand to right-side carry.

His Grabs
    His right hand to your right side.
    His right hand to your left side.
    His left hand to your right side.
    His left hand to your left side.
    He grabs your wrist and hand.
    He grabs your stick.

Positional Counter-Options
    You punch and/or kick the attacker.
    Power body turns.
    You let go of your stick handle and lift your hand. Now it is just a grab.
    You circle the wrist clock-wise or counter-clockwise with your pommel.
    You push-pull release the grab.
    Arm wrap a grab right on the stick handle.

Stick versus Stick Disarm Summary

Single Stick versus stick encounters occur in a few possibilities:
    Possibility 1: SMS vs. SMS Encounters
    Possibility 2: DMS vs. SMS Encounters
    Possibility 3: DMS vs. DMS Encounters
    Possibility 4: Hand Grip Variables
        a) SMS - right vs left hand issues
        b) Hand grips on the stick
            - SMS: one hand on one end
            - DMS: two hand riot grip
            - SMS: one hand center grip
            - DMS: baseball grip

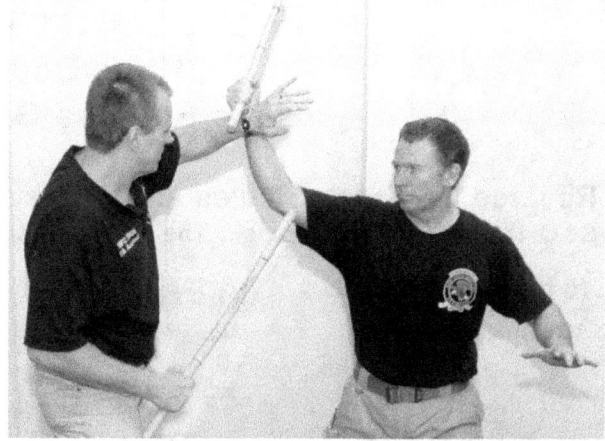

Anatomy of an SMS Attack:
    Shaft attack
    Tip attack
    Pommel attack
    Right hand
    Left hand

Anatomy of a DMS Attack:
    Shaft
    End strikes with pommel/tip
    Basic training on the clock corners

Anatomy of a Block:
    SMS Blocks - Supported and Unsupported
    DMS Blocks

The Big 5 SMS Disarms:
    Impact Disarm
    Hand Snake Disarm
    Stick Snake Disarm
    Strip and Keep Disarm
    Strip and Send Disarm

The Big 3 DMS Disarms:
The lack of a free empty hand disallows hand snakes and strip and keeps.
    Impact Disarm
    Stick Snake Disarm
    Strip and Send Disarm

Practice encounters on the 4 basic clock quadrants.
Practice encounters on the SMS 12 or DMS 15 angles.

Counter disarms with:
    1) Explosive Retraction (and good body position)
    2) Hit the Disarmer
    3) Kick the Disarmer
    4) Hand Switch
    5) Slap Release
    6) Handle Punch
    7) Row and Roll Releases
    8) Charge in and blast away
    9) Unique solutions

CQCG

KNIFE COUNTER/KNIFE COMBATIVES
Knife/Counter-Knife Level 5:
The Reverse Grip Stab Assault Module

Reverse Grip Stab Studies and Observations
S&O 1) The Mad Rush Attack: The Common 10 thru 2 Attack

The common attacker will typically hold a reverse grip and attack in an overhead manner, from a 10 o'clock, through 2 o'clock positions.

The trained fighter may well attack on other lesser instinctive angles of the clock. But, practitioners worried and wishing to survive the impulsive, common knife attacker should start training counters against this range variance first, then the others.

S&O 2) The Reverse Grip is a natural for ground fighting.

It is natural to want to hold a knife in a reverse grip for many ground fighting positions. Experts practice finding a safe and secure moment to flip or fan the blade (see prior *Training Mission* Books) into the grip they require.

The reverse grip holder should also be aware that the knife held in this position may be forced back into his body by the enemy, or he may accidentally stab himself in the grappling scuffle.

In the big picture, there is simply no such thing as one overall perfect knife grip, only that perfect grip for the situation. It is important to study the saber and reverse grip.

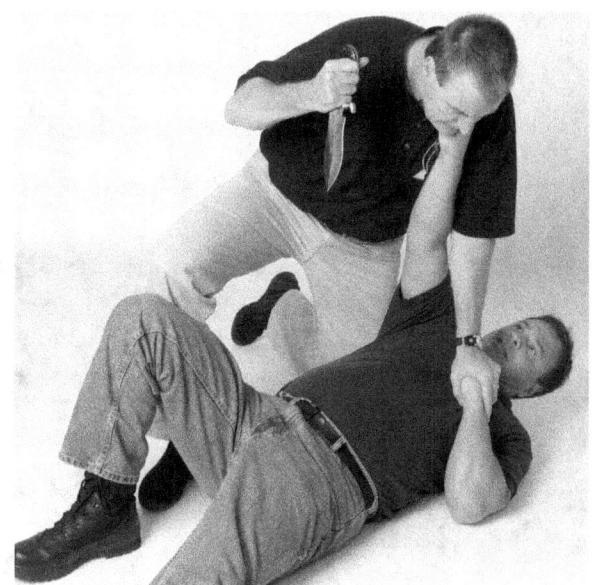

S&O 3) The need for a guard/hilt for stabbing.
The guard or hilt protects your hand from his edged weapon attack when weapons clash.

The hilt, guard or cross guard above the handle protects you from the edge assault of the enemies weapon.

This plate on the left from **Codex Wallerstein -** a European medieval fighting book from the 15th Century shows the long, historical importance of protecting the hands during edged weapon battle.

As this knife course now officially breaches into the subject of stabbing, it is important to note that the knife guard also protects your hand from running up the handle and onto the blade of the weapon. When you produce the force of a thrusting stab, and you hit your target, the knife slows and may stop. With an improper knife, your hand may slip up on the blade.

This is not such a problem with fixed blade knives, as they seem to have a suitable hilt. It is more of a danger with folding knives. Their grips are often designed to be smooth and attractive. A practical, workable folder has great texture on the handle and a shape that is conducive to maintaining a hand grip.

Page 119 - W. Hock Hochheim's Training Mission Five

S&O 4) Reverse Grip Blade Evacuation

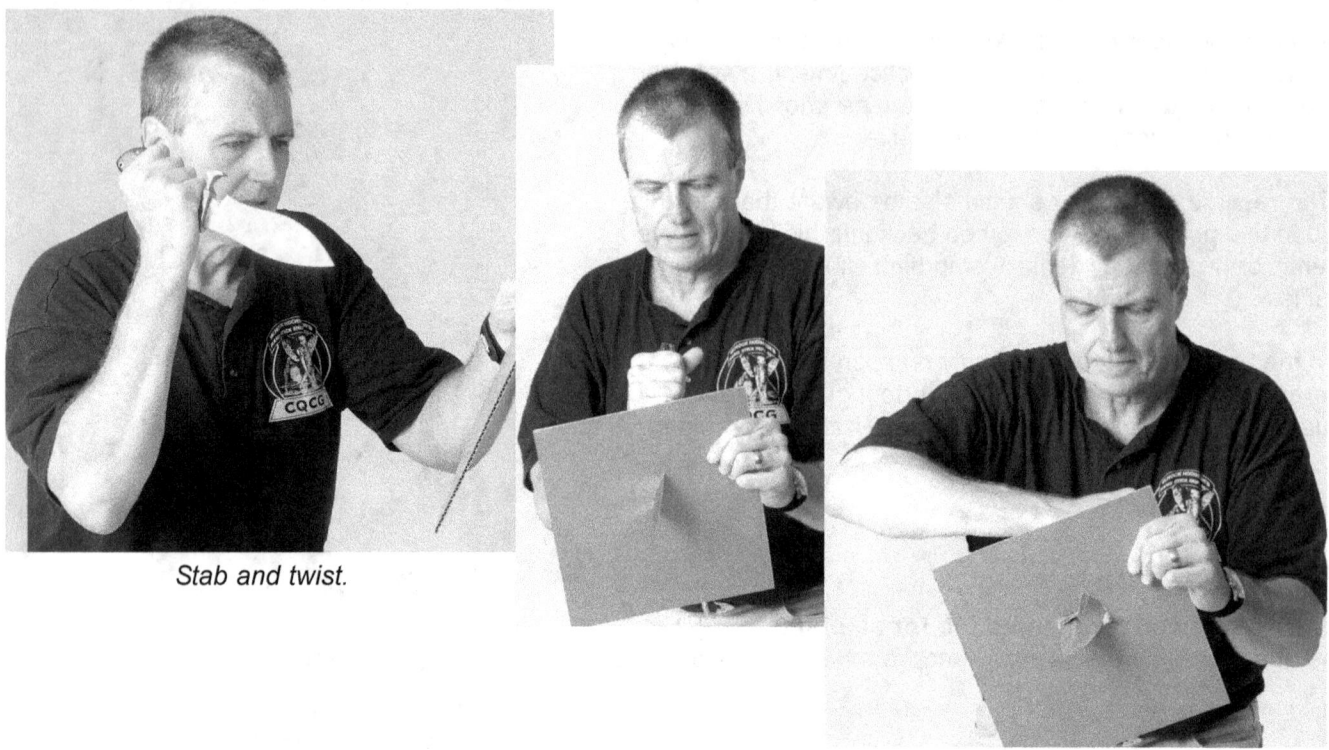

Stab and twist.

A stabbing knife may embed deeply into the bones and flesh of the enemy, and limit your follow-up action. At times people have lost their knives this way, buried in the attacker as the enemy falls down or runs off with your knife in him. This has been remedied through the ages by learning how to twist the blade in any or all of the attack times:

    Possibility 1) Twist during the thrust.
    Possibility 2) Twist during the stab.
    Possibility 3) Twist during the extraction (most common).
    Possibility 4) Move the handle up and down or side-to-side depending upon the stab's entry.

 Twisting the knife opens up the wound canal. It aids blade evacuation. Twist with your wrist, and/or your elbow, and/or your shoulder, and/or your full body. Use one, two, three or all four as needed.

S&O 5) The Wandering Thumb

The wandering thumb issue comes up again as it has in all grip examinations. Place your thumb as needed. Many *cap* off the handle when stabbing for extra reinforcement. The thumb may travel down beside the other fingers to offer solid support in grip strength.

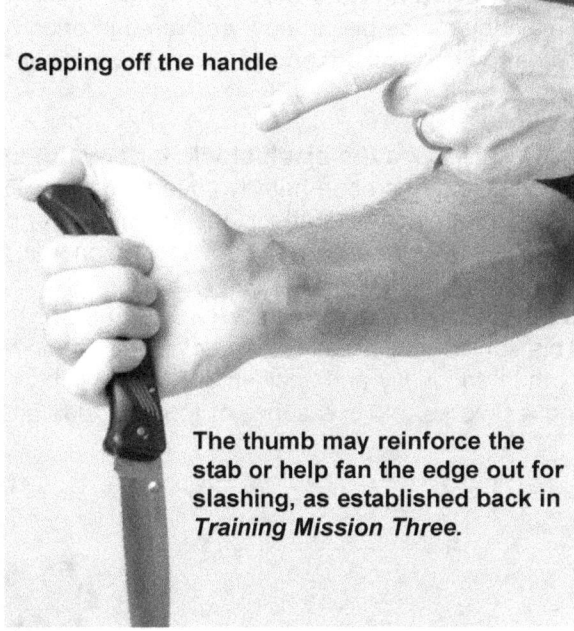

Capping off the handle

The thumb may reinforce the stab or help fan the edge out for slashing, as established back in *Training Mission Three*.

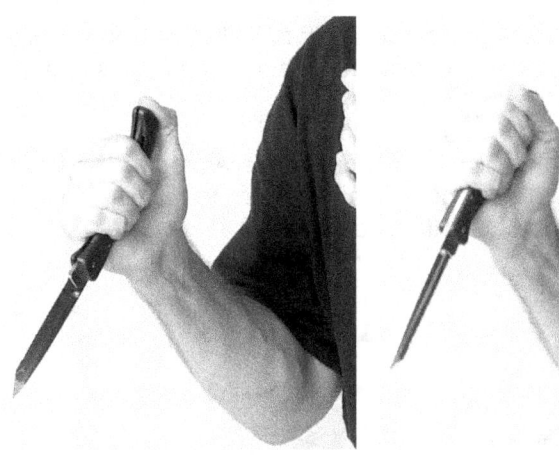

S&O 6) The Support Hand and Legs

Use the free hand. The free hand may grab objects to fight with, always a primary strategy. This free hand strikes, grabs, blocks and confuses. Unique to this reverse grip stabbing study is the use of the free hand to push the pommel of the knife and generate more stabbing power. This tactic is common in Indonesia, India and several other national fighting systems but, as with all techniques, it is probably taught almost everywhere to some degree.

This potential pommel push keeps a free and wandering hand moving in the window of combat, turning palm-in and-palm out as needed. Remember, when the free hand is turned inward, it protects key blood vessels and muscles. You will see me many times in this knife section with a palm turned outward because the photographer caught me in shadowboxing moments. If the enemy knife draws near, turn your vital vessels and tendons inward.

The free hand is invaluable and must be cultivated in both your solo and partner practice. The eye jab, or any hand strike, is like the "secret" slipped-in shot that often wins the day.

This support weapon concept also holds true for kicking. Always develop your hand-to-hand skills for they are the foundation of weapon fighting.

Pushing, pulling, grabbing and striking

S&O 7) Uncommitted Stab
This subject has been heavily covered in prior *Training Missions*. When you go after a target, remember the path may be blocked, and you may have to attack another target. Be free enough in your attack to change targets.

S&O 8) Beware the slash/block. It does not stop an attack.
This subject has been heavily covered in prior *Training Missions*. If you absentmindedly slash instead of block-and-stop an incoming attack, the incoming weapon will still drive inward as you slash off of it. The attack may only be slowed, and will still advance upon you as you slash off the attack. Looks pretty, but a deadly mistake.

S&O 9) Beware the propensity to self-stab!
This subject has been heavily covered in prior *Training Missions*. In grappling chaos, the reverse grip can be both intentionally and accidentally turned into your body. I will always ask practitioners to learn to fall while holding a reverse grip in a series of forward, side and back rolls, emphasizing knife awareness.

*Practice falling while holding a knife to create edge and tip awareness.
It is easy to stab yourself with a reverse grip in the chaos of combat.*

S&O 9) Success of the Stab? Tenacity of life? Targets?
The general rule of medical experts is a blade must penetrate at least three inches to be fatal. But I'm leery of this generalization. It will always be more about WHERE someone has been stabbed. Anyone old enough to be reading this book, by now, knows places on the body where stabs will be more effective than not.

A central American soldier displays the bloody knife he used in a reverse grip, in Iraq, to kill terrorists in a close quarter battle. The fight was not over quickly.

Reverse Grip, Solo Command and Mastery Drills

The Basic Reverse Grip Stabbing Drills

Drill 1) 4 Angle Hooking Stab Clock Basic Series
* execute right handed and left handed
* execute walking forward, then back
* execute on the ground
* execute standing
* execute kneeling

12 o'clock hooking stab

3 o'clock hooking stab

6 o'clock hooking stab

9 o'clock hooking stab

Drill 2) 4 Angle Thrusting Stab Clock Basic Series
* execute right handed and left handed
* execute walking forward, then back
* execute on the ground
* execute standing
* execute kneeling

12 o'clock thrusting stab

3 o'clock thrusting stab

6 o'clock thrusting stab

9 o'clock thrusting stab

Page 123 - W. Hock Hochheim's Training Mission Five

The 10 Angle Reverse Grip Stabbing Advanced Drill

* execute right handed and left handed   * execute walking forward, then back   * execute on the ground

Page 124 - W. Hock Hochheim's Training Mission Five

From a right hand grip:
Angle 1: High-hooking right stab
Angle 2: High-hooking left stab
Angle 3: Medium-hooking right stab
Angle 4: Medium-hooking left stab
Angle 5: Low-hooking right stab
Angle 6: Low-hooking left stab
Angle 7: Uppercut hooking stab
Angle 8: Downward hooking stab
Angle 9: High straight stab
Angle 10: Low straight stab

From a left hand grip:
Perform the opposite

Walk forward and backward

Perform on your back

Basic Combination Stab Drills

Review! Before introducing combinations involving slashes, it is important to review these stabbing basics and knowledge of a knife after a penetrating stab. A stabbing knife may embed into the enemy and limit your follow-up action. This has been remedied through the ages by learning how to twist the blade in any or all of the attack times;

Possibility 1) Twist during the thrust.
Possibility 2) Twist during the stab.
Possibility 3) Twist during the extraction (most common).
Possibility 4) Pumping knife up or down or side-to-side.

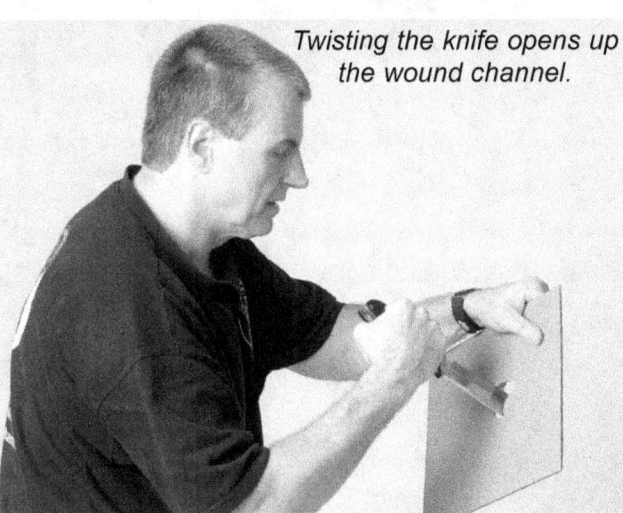

Twisting the knife opens up the wound channel.

It aids blade evacuation.

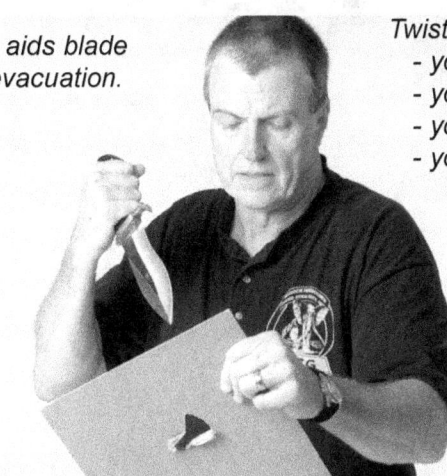

Twist with:
- your wrist
- your elbow
- your shoulder
- your full body

Use one, two, three or all four as needed.

Reverse Grip Stab Combination 1) Stab and slash sets, as performed on the 4 clock angles.

A 12 o'clock stab

1) The 12 o'clock Stab and Slash

Once stabbed, twist and rip the knife across and out in a slashing manner.

A 3 o'clock stab.

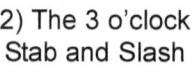

2) The 3 o'clock Stab and Slash

Once stabbed, twist and rip the knife across and out in a slashing manner.

A 6 o'clock backhanded stab.

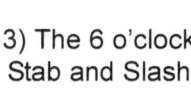

3) The 6 o'clock Stab and Slash

Once stabbed, twist and rip the knife across and out in a slashing manner.

A 9 o'clock backhanded stab.

4) The 9 o'clock Stab and Slash

Once stabbed, twist and rip the knife across and out in a slashing manner.

Exercise these combinations left handed.

Reverse Grip Stab Combination 2) Slash and Stab sets, as performed on the 4 clock angles.

Reverse Grip, Solo Command and Mastery Drills

The Basic Reverse Grip Stabbing Drills

Drill 1) 4 Angle Hooking Stab Clock Basic Series

1) The 12 o'clock Slash and Stab

* execute right handed and left handed
* execute standing
* execute walking forward, then back
* execute kneeling
* execute on the ground

A 12 o'clock downward slash.

At the most economic point, you stab.

2) The 3 o'clock Slash and Stab

A 3 o'clock horizontal slash.

At the most economic point, you stab.

3) The 6 o'clock Slash and Stab

A 6 o'clock upward slash.

At the most economic point, you stab.

4) The 9 o'clock slash and stab.

A 9 o'clock backhand, horizontal slash. At the most economic point, you stab.

Exercise these combinations left handed.

Reverse Grip Stab Combination 1) Stab and hand strike sets, as performed on the 4 clock angles.

Use any hand strike inside the clock formation.
    Occurrence a) Hand strike and stab
    Occurrence b) Stab and hand strike

Practice:
    Clock 4 with stab and hand strike sets

    Clock 4 with hand strike and stab sets

    Do with the above opposite hand grip

    Do while walking forward and back

    Do the above on your back

    The 12 Angles of Attack combinations

Reverse Grip Stab Combination 2) Stab and kick sets, as performed on the 4 clock angles.

Use any kick inside the clock formation.
    Occurrence a) Kick and stab
    Occurrence b) Stab and kick

Practice:
    Clock 4 with stab and kick sets

    Clock 4 with kick and stab sets

    Do with the above opposite hand grip

    Do while walking forward and back

    Do the above on your back

    The 12 Angles of Attack combinations

Reverse Grip Knee-High Maneuver Drill

By now, you should be familiar with this knee-high drill, one I first created specifically to combat the dangers of the reverse grip and its propensity to self stab. Here again, we remind and practice the knee-high drill.

The Floor Clock Moves:

1) Center axis
2) Right knee to 12 o'clock
3) Center axis
4) Left knee to 12 o'clock
5) Center axis
6) Right step to 1:30
7) Center axis
8) Left step to 10:30
9) Center axis

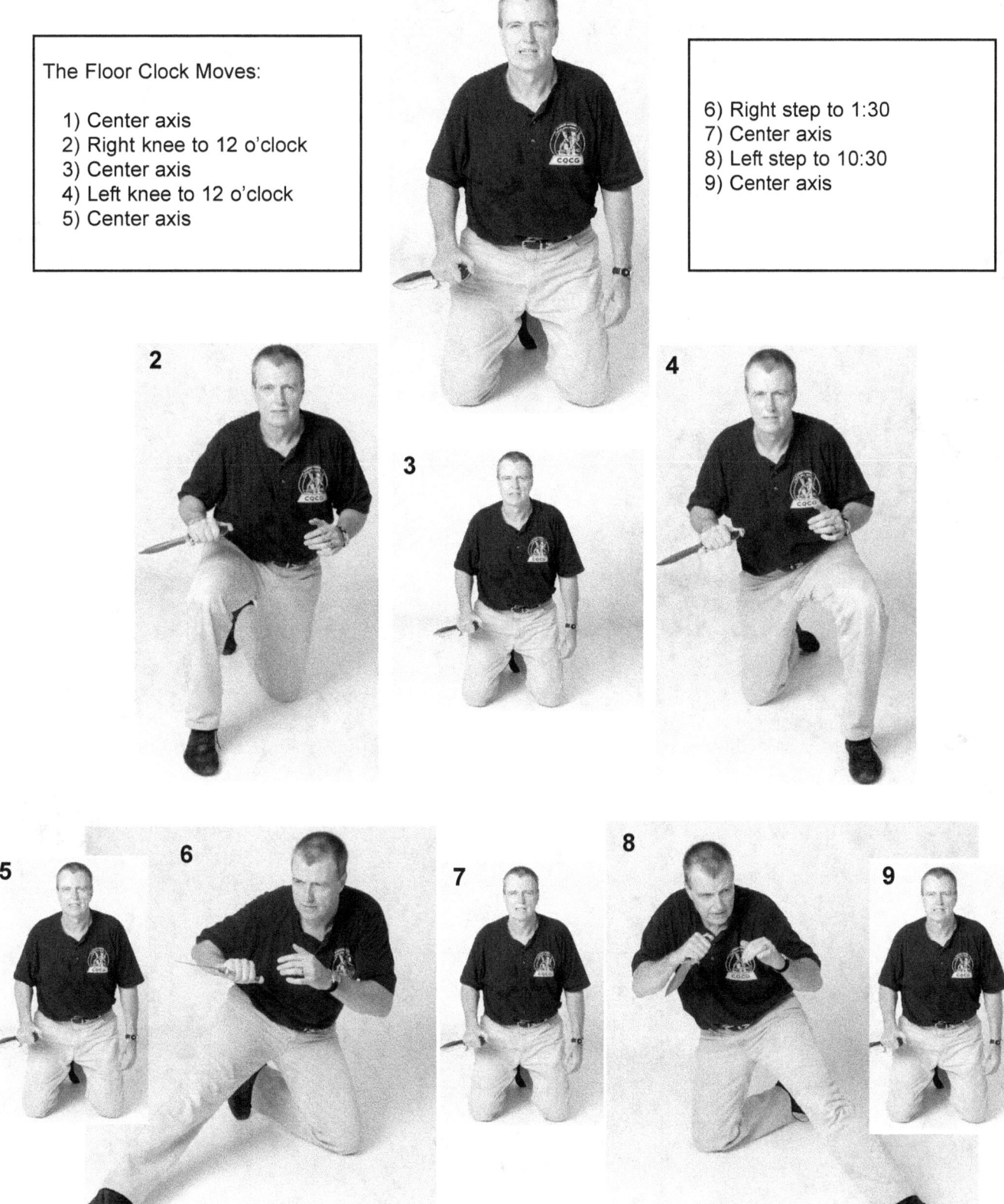

Page 131 - W. Hock Hochheim's Training Mission Five

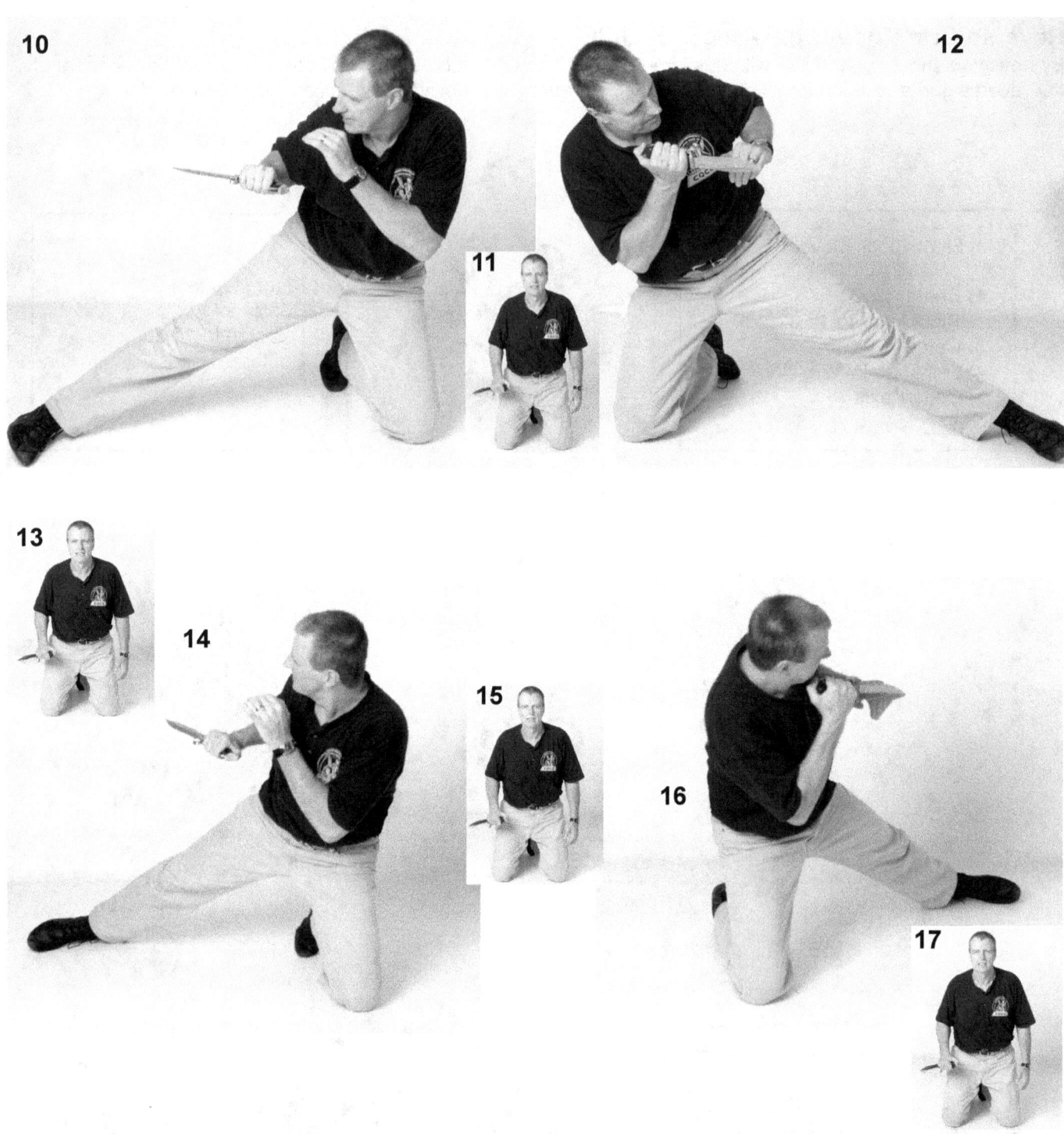

10) Right leg out to 3 o'clock
11) Center axis
12) Left leg out to 9 o'clock
13) Center axis
14) Right leg out to 4:30
15) Center axis
16) Left leg out to 7:30
17) Center axis

18

18) Center axis

19) Right leg to 6 o'clock

19

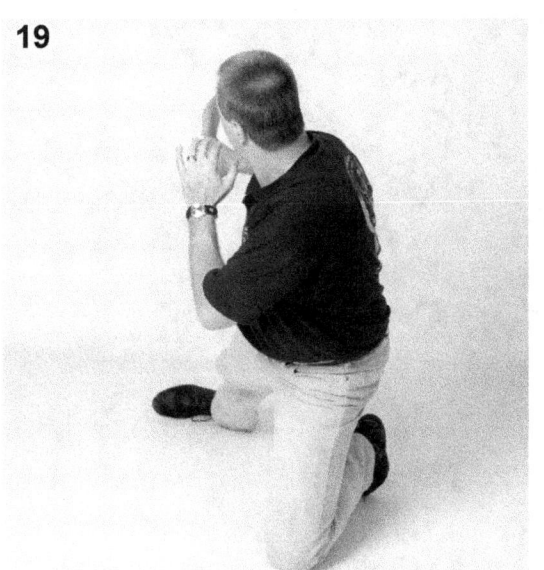

20

21

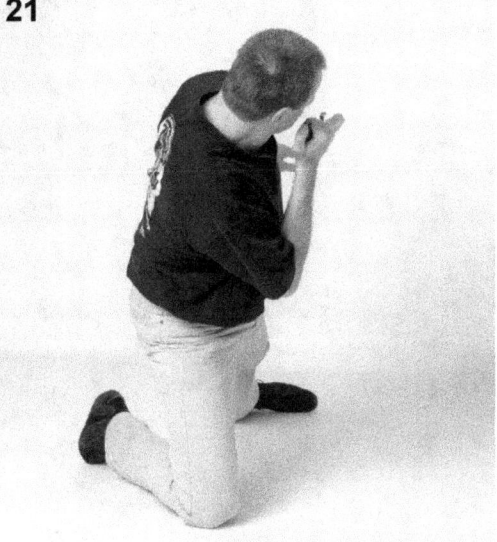

22

20) Center axis
21) Left leg to 6 o'clock
22) Center axis

Hit Training Objects

Once you have graduated from solo "in the air" practice, next you must take your reverse grip to a wooden post or any striking object and work the 4 and 12 angle drills with hard contact.

Stationary Post Training

Next, a training partner is armed with training devices. The trainer prepares to coach the trainee. The stick is for him to slash. The small shield is for him to stab.

The trainer uses a pad or shield as a target to strike. Keep the interaction lively and cover some ground.

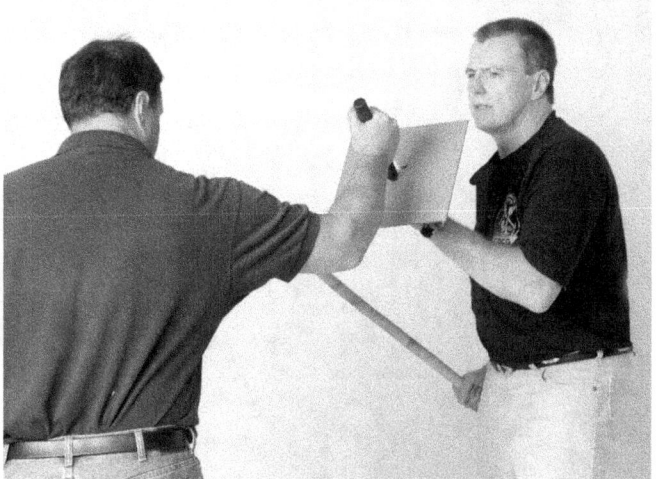

The trainer uses a stick to elicit a reverse grip slash.
Then a flat target to draw a stab.

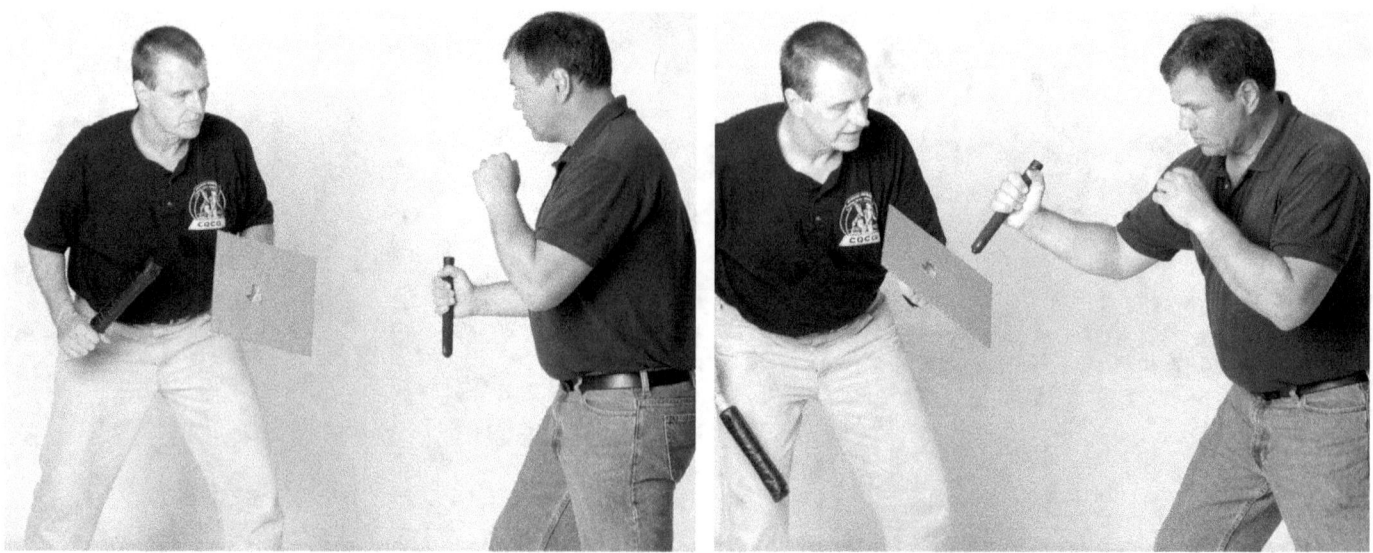

Next, the trainer strikes back! This increases the speed of the trainee's attack, footwork and responses.

Next, the trainer starts striking back with a training tool. The trainee must improve his footwork to strike and escape, or at least strike and strike/block the incoming attack. This is the progression step right before full sparring. Most practitioners realize quickly that knife dueling is best done with a saber grip. The range advantage is far superior than sparring with a reverse grip.

Still, there are practitioners who prefer to do all knife fighting with a reverse grip. Some even prefer the sharp edge in and the dull edge side, or half-sharp side, facing the enemy. Their plan is to stab you, of course, but also to hook you in and cut your arm in the hooking process. I have been deeply involved in knife training and have taught thousands of people for well other many years. I watch, test and evaluate. But, if that is not enough, your own freestyle experimentation will prove that saber grip dueling has an advantage.

First, getting near enough to a moving duelist with a reverse grip is difficult. Secondly, hooking your opponent, if you indeed do get near enough to catch him and hook his arm, pulls he and his blade right into you. In many ways pulling an armed opponent into you like this is simply suicide.

No matter the statistics and the results, these stylists will argue for and prefer the reverse grip for all encounters, all the time.

The Basic Saber Stab Statue Drill

The "Outside / Inside / Split / Inside / Outside" statue progression. The trainer uses his empty arms and hands for your skill development as shown in the prior *Training Mission* books. This is a rudimentary training drill, teaching the novice the ways and means of the knife as it interacts with a body. It is an excellent teaching technique. The three formats remain the same as in the prior levels:

Format 1) Knife makes arm contact / knife stabs neck (or any target).
Format 2) Hand makes arm contact / knife stabs neck (or any target).
Format 3) Double: hand and knife makes arm contact, knife stabs neck (or any target).

FORMAT 1

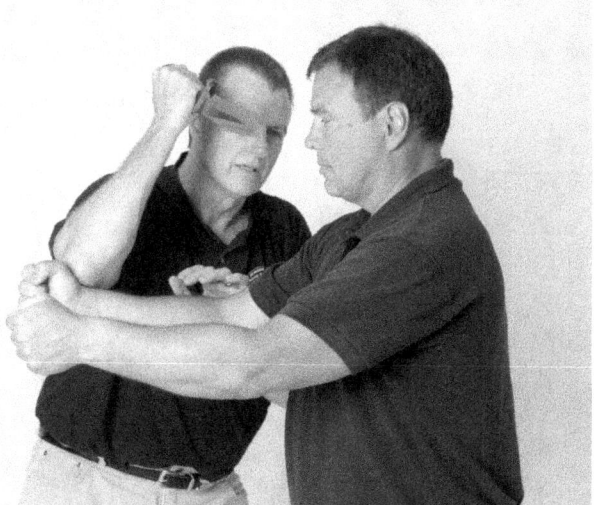

Outside the arm, knife-to-arm contact, then knife stab to throat. My left hand is up for cover and control.

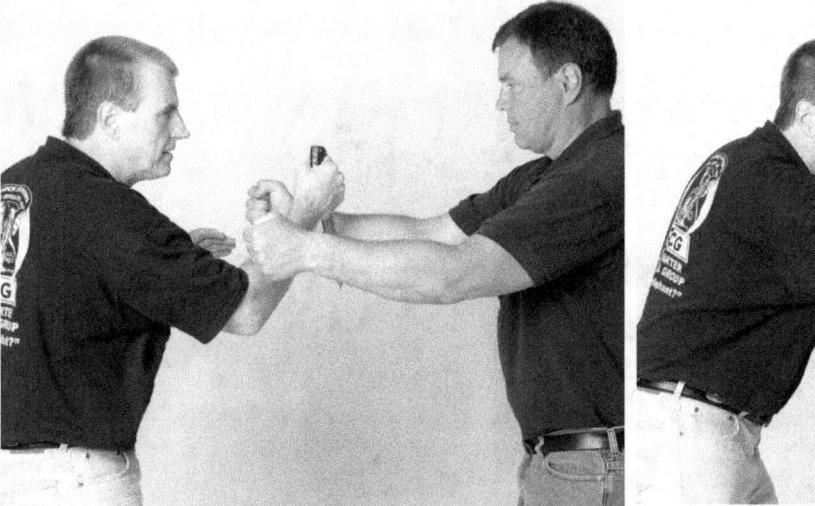

Inside the arm, knife-to-arm contact, then knife stab to the throat. My left hand is up for cover.

FORMAT 1 cont...

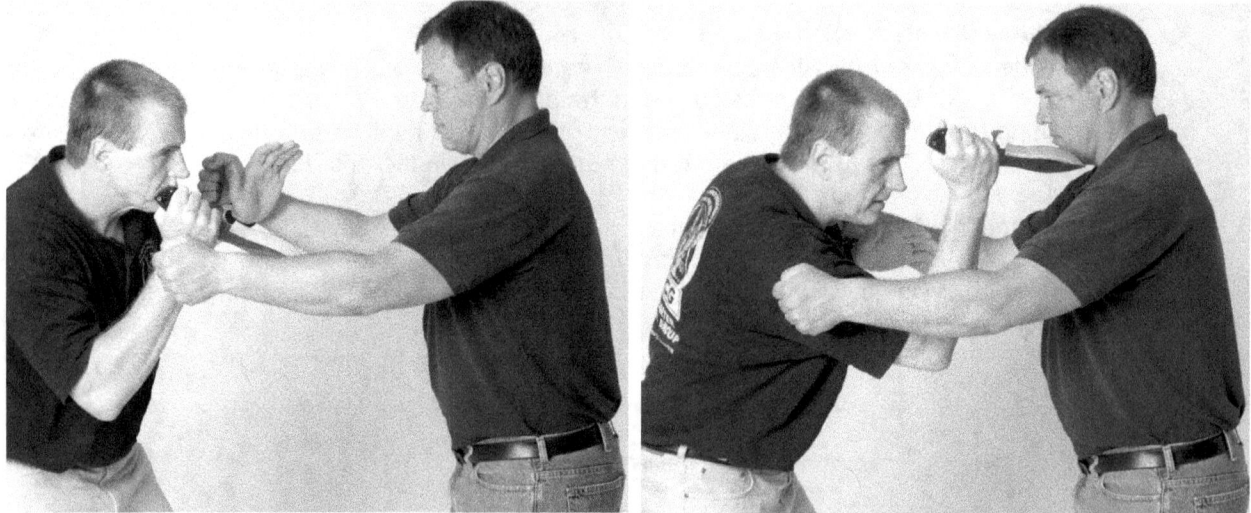

Split! Block both arms. Inside the arm and knife-to-arm contact, then knife stab to throat.

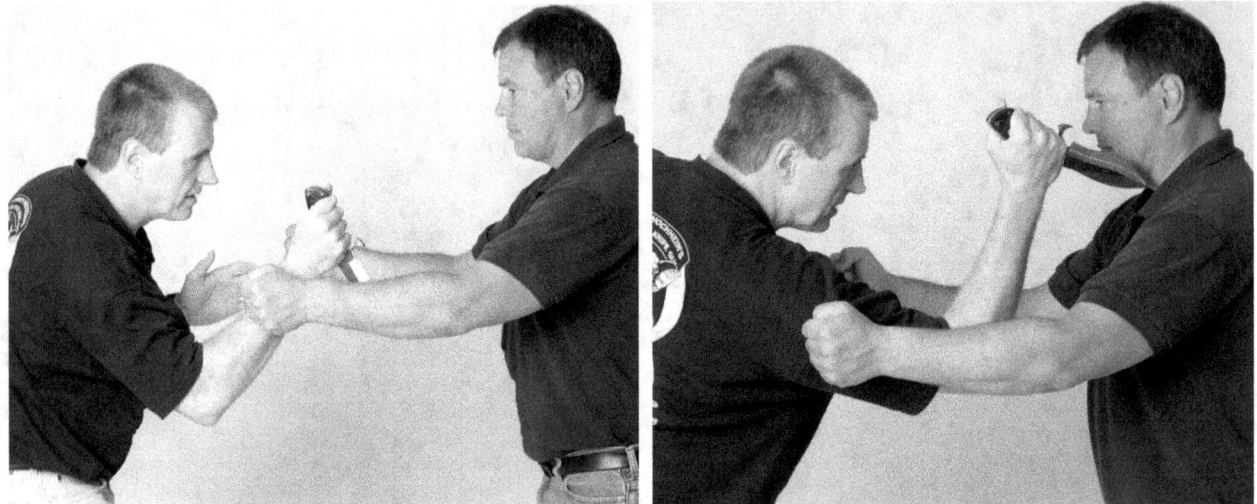

Inside his left arm contact, then knife stab.

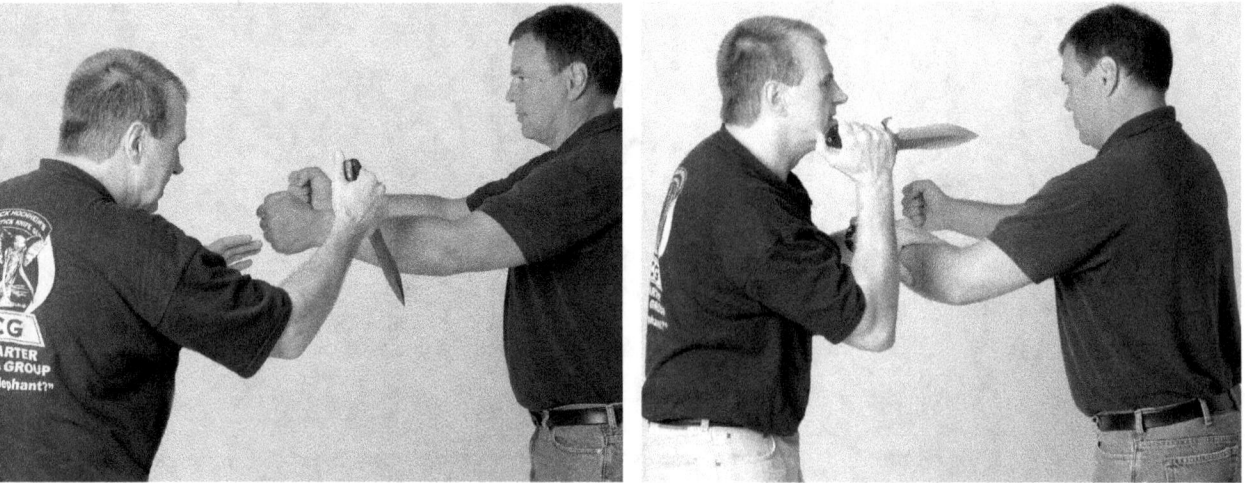

Outside contact. My left hand comes up for cover. The knife stabs.

FORMAT 2

Hand makes the first contact on the outside right arm, and immediately stab with the knife.

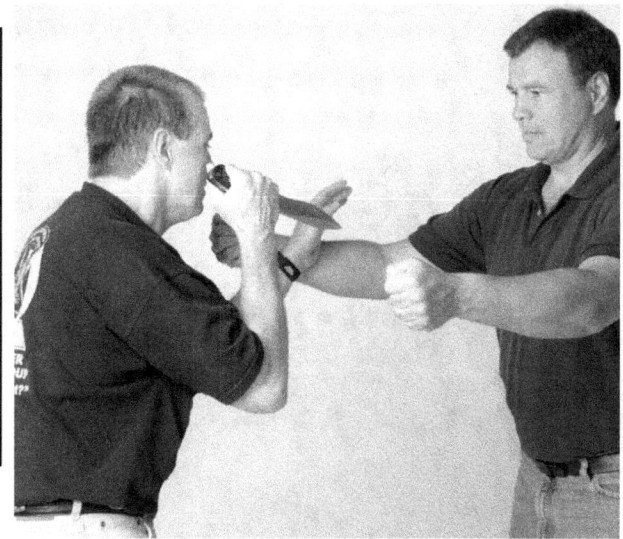 

Hand makes the first contact on the inside right arm, and immediately stab with the knife.

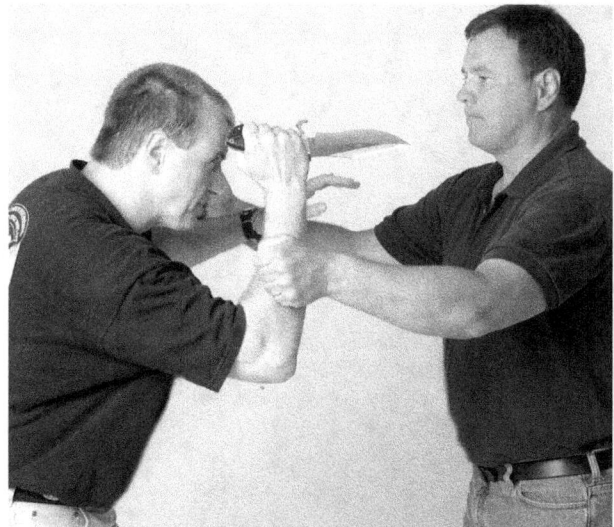 

Split! Block both arms. Inside the arm and knife-to-arm contact, then knife stab to throat.

FORMAT 2 cont...

 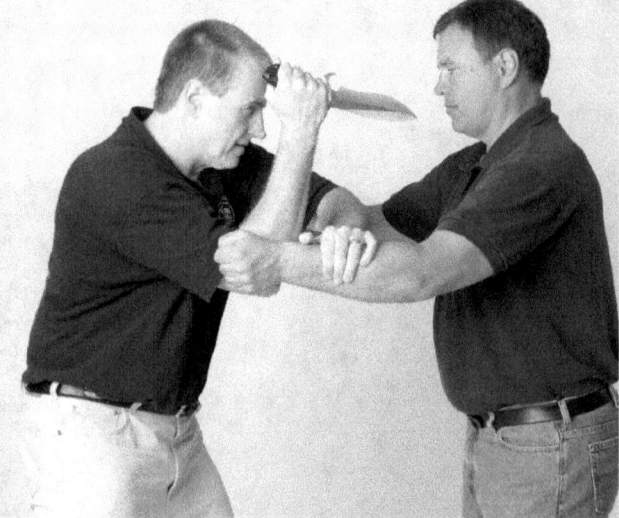

Hand makes the first contact on the inside left arm, and immediately stab with the knife.

 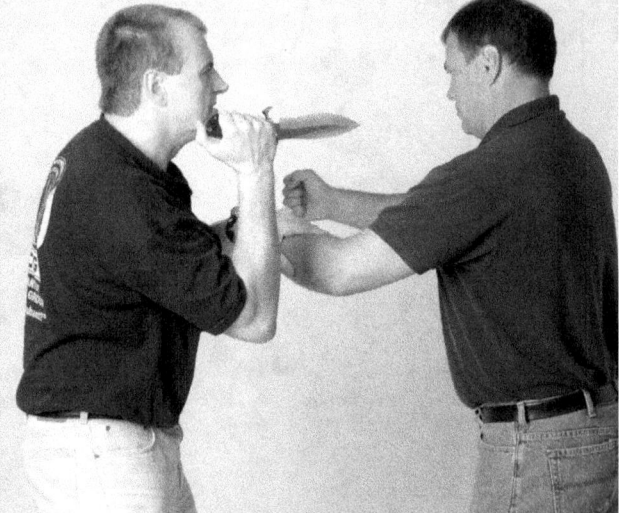

Hand makes the first contact on the outside left arm, and immediately stab with the knife.

FORMAT 3

Practice a third format, where both the knife and the empty arm makes a simultaneous contact on the trainer's arm, then the knife attacks with a stab. The empty hand offers cover.

Continue to invent practical formats.

Reverse Grip Stab Counters to Common Blocks Study:

By now, you should have a working knowledge of this classic drill. It does nothing but develop your knife manipulation skills in an increasing progression. As shown in the prior *Training Mission* books, this is the rudimentary training drill, teaching the novice the ways and means of stabbing as it interacts with basic blocking. It is an excellent teaching technique. The trainer uses his empty arms and hands for your skill development. These are not combat scenarios. The three formats remain the same as in the prior levels. The trainer will block them with his arm. You will defeat the arm in three formats.

Format 1) Cut the block.
Format 2) Re-direct attack on another line.
Format 3) Invading hands.

The Four Basic Clock Angle Attacks

Your 12 o'clock: The first is a high attack.
He blocks high with his right or left arm.

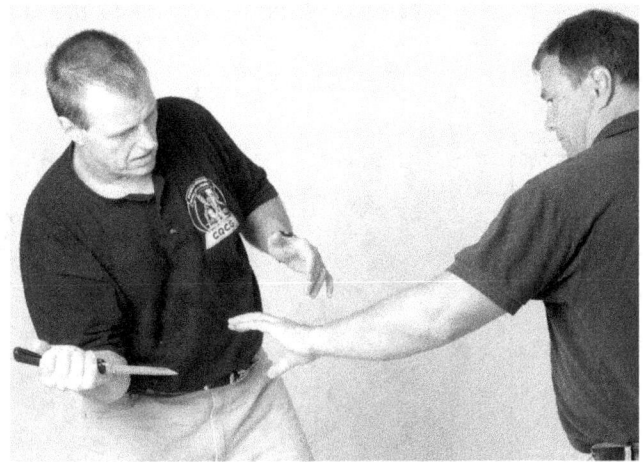

Your 3 o'clock. You attack from the right.
He blocks to his left, with his right or left arm.

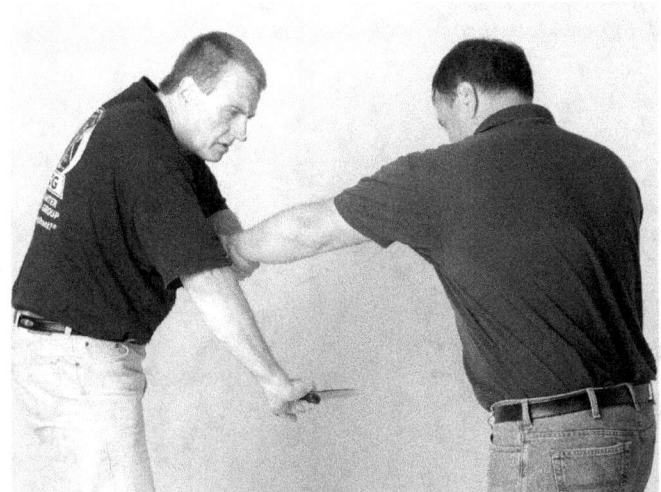

Your 6 o'clock. You attack from low.
He blocks low with his right or left arm.

Your 9 o'clock. You attack from the left.
He blocks to his right with either arm.

FORMAT 1

 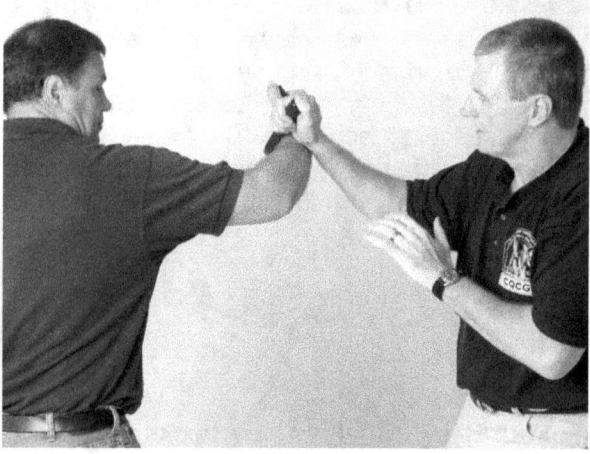

Cut the Block Series. Your attack is thwarted. You cut the blocking limb anyway you can.
As shown in the second photo, this is the common counter. Bring in the edge to cut and yank back.
Work through the 4 clock angle progression by cutting the blocking limb.

FORMAT 2

 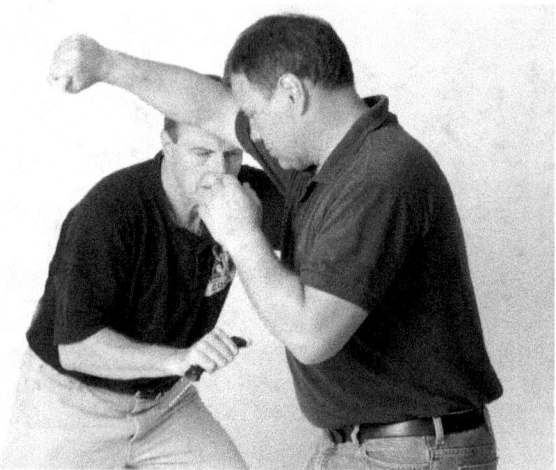

Re-direct Series. You attack on the clock angles. When thwarted, you re-direct your attack on another open line. At times the opening may call for a slash. At times a stab.
Use you free hand for cover. Work through the 4 clock angle progression.

FORMAT 3

 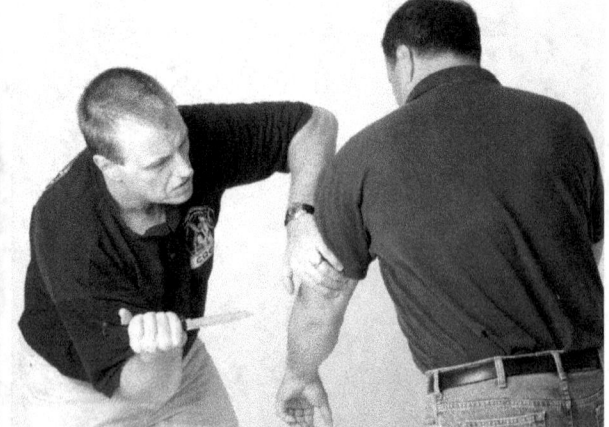

Invading Hands Series. You use your free hand on the opponent's limbs to clear a path to a better target. Pin. Pass. Pull. Push. At times the opening may call for a slash. At times a stab.
Work through the 4 clock angle progression.

Your Counters to Common Blocks Workout

1) Cut the block vs. an unarmed training partner.
2) Re-direct on another line of attack vs. an unarmed training partner.
3) Invading hands vs. unarmed training partner, using the Four P's:

   Pinning hands
   Passing hands
   Pulling hands
   Pushing hands

4) Counter 1 block - one layer of obstruction.
5) Counter 2 blocks - or two layers of obstruction.
6) Do all of the above with the knife in your other hand.
7) Exercise some of these from ground positions.
8) Put a knife in the trainer's hand and experiment with these situations. As soon as you arm the training partner, these drills start looking like complicated combat scenarios. Remember! This set of exercises are just simple training, skill drills to develop knife manipulation skills. They are steps to be inserted into combat scenarios later.

Reverse Grip Synergy Flow and Skill Drills

Before, we developed "dummy" drills against an armed partner. The trainer portrayed a moving dummy by simply tossing his arms and hands in the way to develop obstacle removal. In the next series of drills, the trainer holds a knife, and his involvement escalates. These are not combat scenarios yet, just skill sets and building blocks for them.

Drill 1) The Reverse Grip, Stabbing Block-Pass-Pin Skill Drill.

The trainer uses his empty arms and hands for your skill development. At this level of training, you may also put a knife in his hand as a prop to be concerned about. These are not combat scenarios! As shown in the prior *Training Mission* books, this is the rudimentary training drill, teaching the novice the ways and means of stabbing as it interacts with a body. It is an excellent teaching technique. Practice them:

   a) standing: stabs on the half-beat
      - right hand   - left hand

   b) ground: stabs on the half-beat
      - right hand   - left hand

You know this drill. Six beats. You block, pass and pin. He blocks, passes and pins.
This is the skeleton for the skill developing inserts.

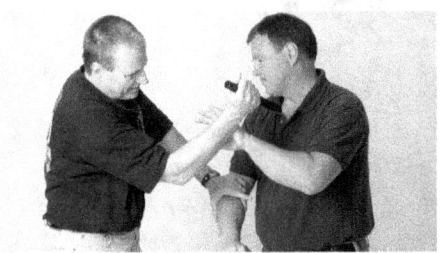  

BEAT 1 1/2 INSERT

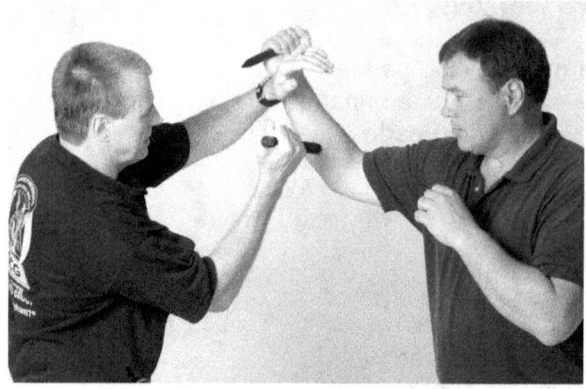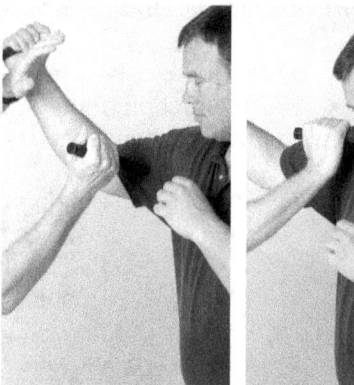

On beat one and a half, stab the lower arm. Next series-upper arm. Next series-throat.

BEAT 2 1/2 INSERT

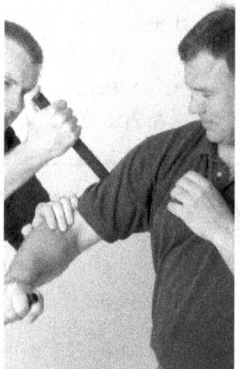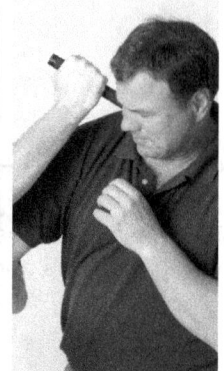

On beat two and a half, stab the lower arm. Next series-upper arm. Next series-throat.

BEAT 3 1/2 INSERT

On beat three-and-half, stab the lower arm. Next series-upper arm. Next series-throat.

BEAT 4, 5, 6 1/2 INSERTS

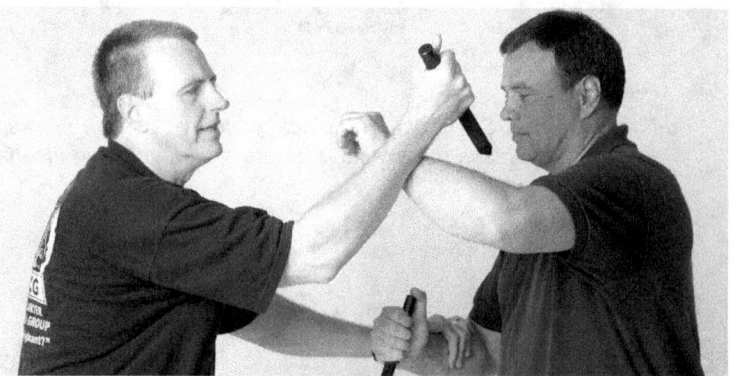

Experiment with stabs on beats 4, 5, 6. Hitting the lower and upper arm, and neck in succession will be blocked on this side. You will find that the previous counters to common blocks offer many more solutions.

O
T
H
E
R

H
A
N
D

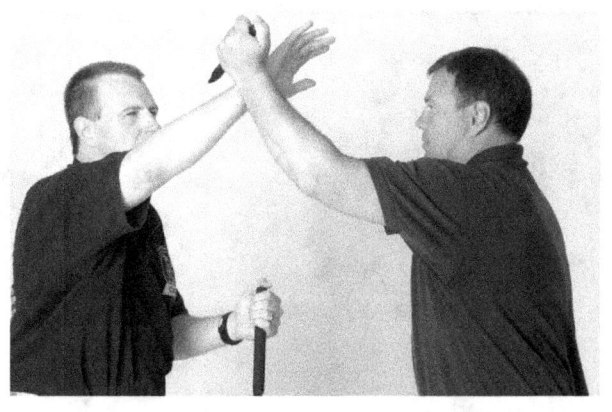
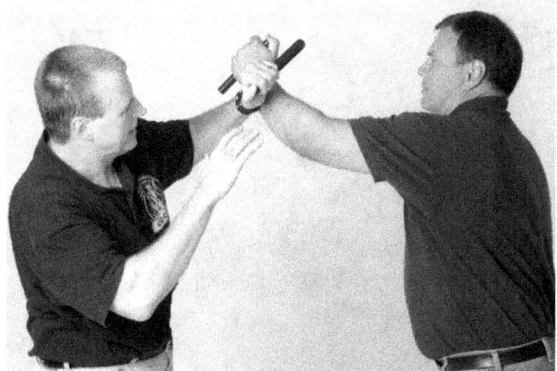

Now work these same target acquisition drills with your weak-side hand holding the knife.

Build fantastic coordination, awareness, savvy and skill with these performance drills.

Ground Block, Pass and Pin

Run this block, pass and pin drill versus leg maneuvers. You learn savvy to work against a downed opponent. He also learns to utilize his legs. You stab in the half beats. Lower leg. Upper leg. Groin.

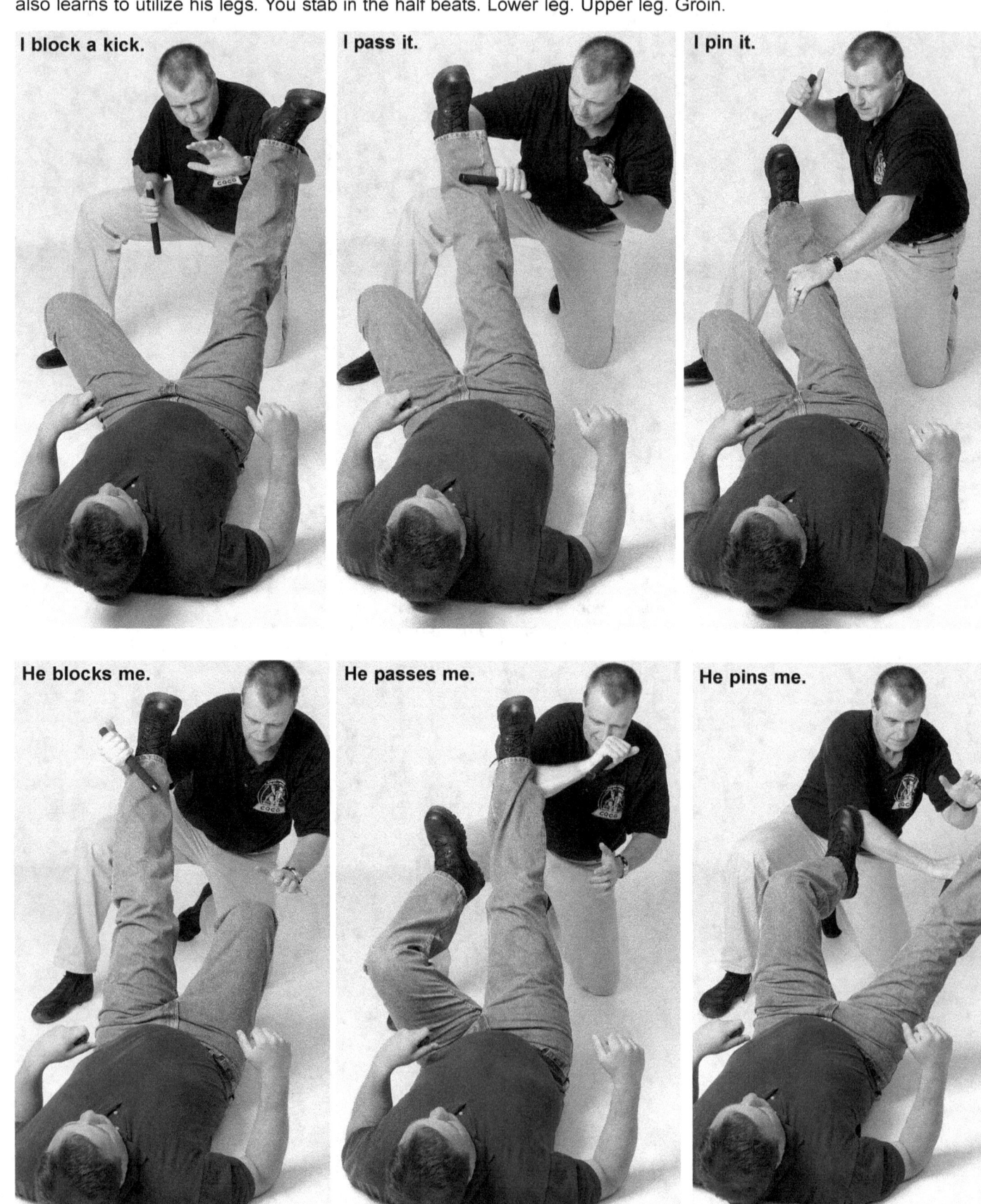

I block a kick. | I pass it. | I pin it.
He blocks me. | He passes me. | He pins me.

Drill 2) The Basic Windmill Drill

Simple and pure in design and motion, yet full of skill development, the windmill drill attacks with the common downward stab. The stab is dodged with a sidestep, while hooked with a cupped hand, then passed aside. Inserts are classified as high, medium or low responses, or early, mid or late phase. The basic motions are just down in two steps or beats.

I stab down. He passes my attack.

He stabs down. I pass the attack.

The attack is pushed at the bottom.

A sample of the common hand cup. There will be many different grips and contacts through this drill.

The limb-to-limb contact potentials:

1) cupped hand passes
2) hand grabs thumb-up
3) hand grabs thumb-down
4) top of your forearm to the bottom of his forearm
5) bottom of your forearm to the top of his forearm
6) your knife crosses over and hooks his knife limb

High Insert Sample:
This happens in approximately the first 15 percent of the downward stroke.

I do not pass but rather interrupt the motions and stop the attack at his high point. If this stop isn't high enough, then you cannot do this move. Make a quick distracting and diminishing stab to his face or neck. If his hand gets in the way, stab wherever you find an opening, but you cannot linger in this position.

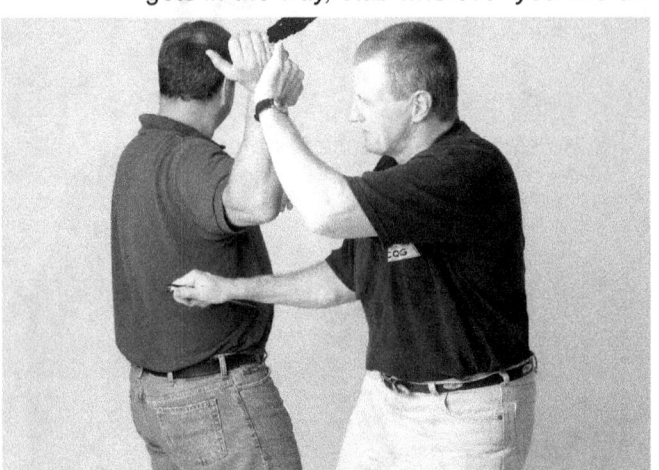 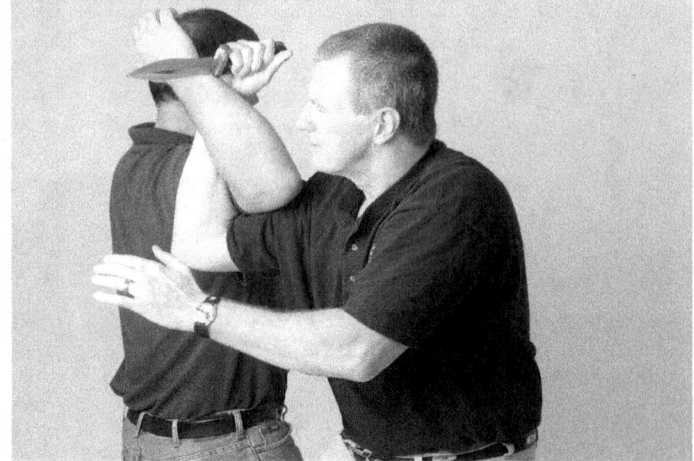

*Charge in and cut if the opportunity is available. I hook the arm and execute an underarm takedown as detailed in **Training Mission Four** book and DVD set.*

In this insert shown here, I do not pass the downward stab, but rather interrupt the windmill motions and stop the attack at his high point. If the stab isn't high enough, then you cannot do this move. Your forearm stops him under his arm in a very reflexive manner.

Stop it and make a quick stab to his face or neck. If his hand gets in the way, stab wherever you find an opening, but you cannot linger in this position. Charge in and cut if the opportunity is available.

Remember this cut may have to penetrate clothing and may not have and immediate effect on your enemy.

Here, I hook the arm and execute an underarm takedown as detailed in **Training Mission Four** book and DVD set.

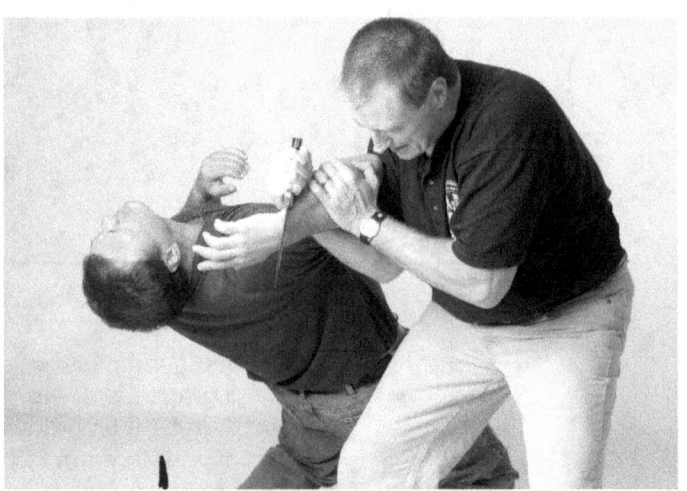

The underarm takedown.

High Insert Sample: A Snake Disarm

This happens in approximately the first 15 percent of the downward stroke. This snaking action is worthy of mentioning at this point because I have heard numerous accounts from military and police personnel that they executed this disarm. Plus, I have investigated cases where untrained criminals have disarmed weapons of their victims and taken flashlights and batons from police officers. The maneuver is often tagged as martial artsy by the ignorant.

 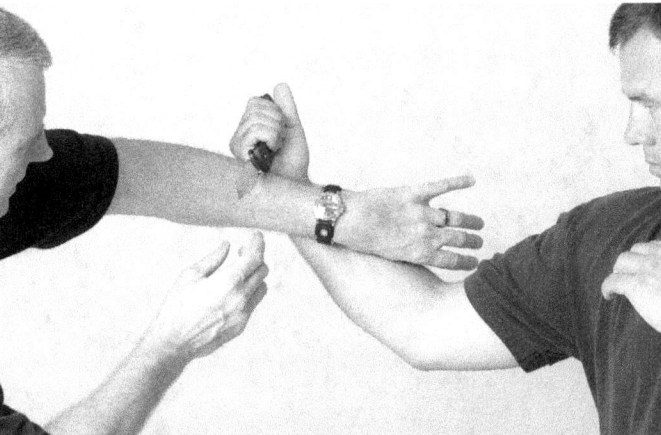

The high reflexive stop. The elbow has a high angled cant. Look -- no thumb on the snake.

 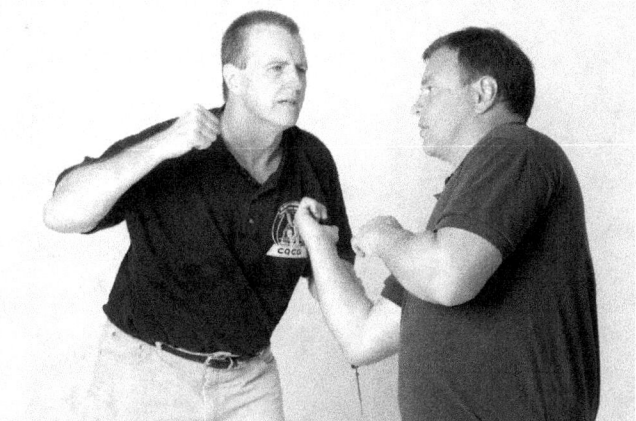

Snake the arm. The knife drops out as you punch him in the process. Some people suggest using two hands to work on the weapon-arm, but this leaves you exposed to his free hand.

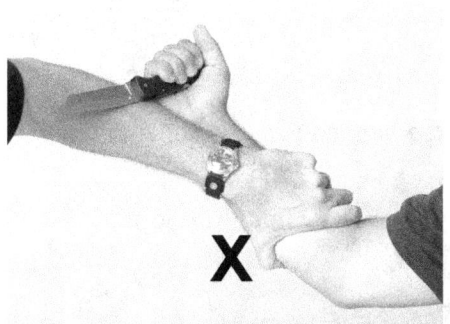

X marks the thumb spot! Many novices grasp the forearm. This stops the snaking process. "Hide" the thumb. Tuck it away instead, and snake the whole forearm.

With a minimum of muscle memory work someone can master this move. Your forearm actually goes up to the blade, which frightens many. I have experimented doing this with live blades and have never been cut. Not much of a problem with kitchen knives and knives have half an edge on the edge. Double-edge knives might cut your forearm.

Sharp-edge facing outward is the aggressive position of choice when the enemy brandishes a weapon. There is a minority of knife stylists who advocate edge-in, but they cannot convince the rest of us to present a weapon in this limited manner. Statistics show that criminals use single-edge kitchen knives up to 98 percent of the time as their weapon of choice. The next type of knives is the half-sharp back edge. Your forearm works past the sharp edge, if the edge is outward.

Versus a double-edge knife? Or one held with a sharp-edge inward? My *Minimization of Wounds Theory* - that it is better to get a slice on the arm than a stab in the neck must dictate out action.

I ask a novice to try this under the guidance of an expert. The veteran can immediately fix the common mistakes. Once you get this movement, you will begin to respect it.

Consider also that this snaking may not be the first event in a fight and would work considerably better against a stunned person.

Medium Insert Sample: An Arm Bar Hammerlock
This happens approximately in the middle of the downward stroke.

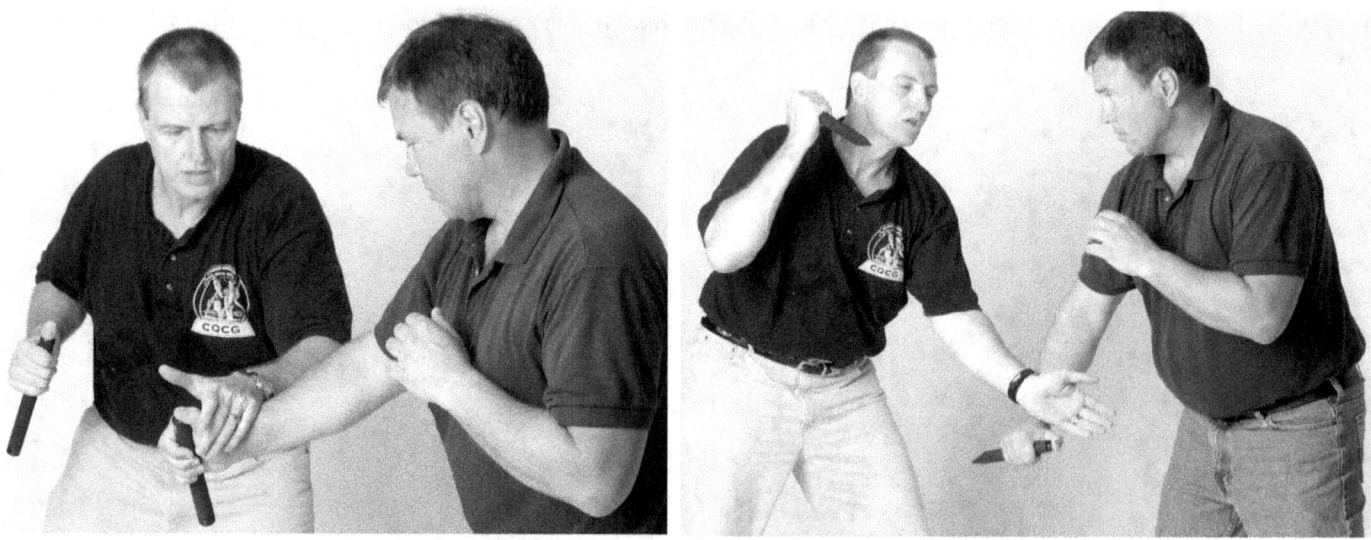

Review the arm bar hammer lock basics from **Training Mission Four**. The bottom of your arm pushes the top of his arm. After practice you will be able to better judge if you can do this or not.

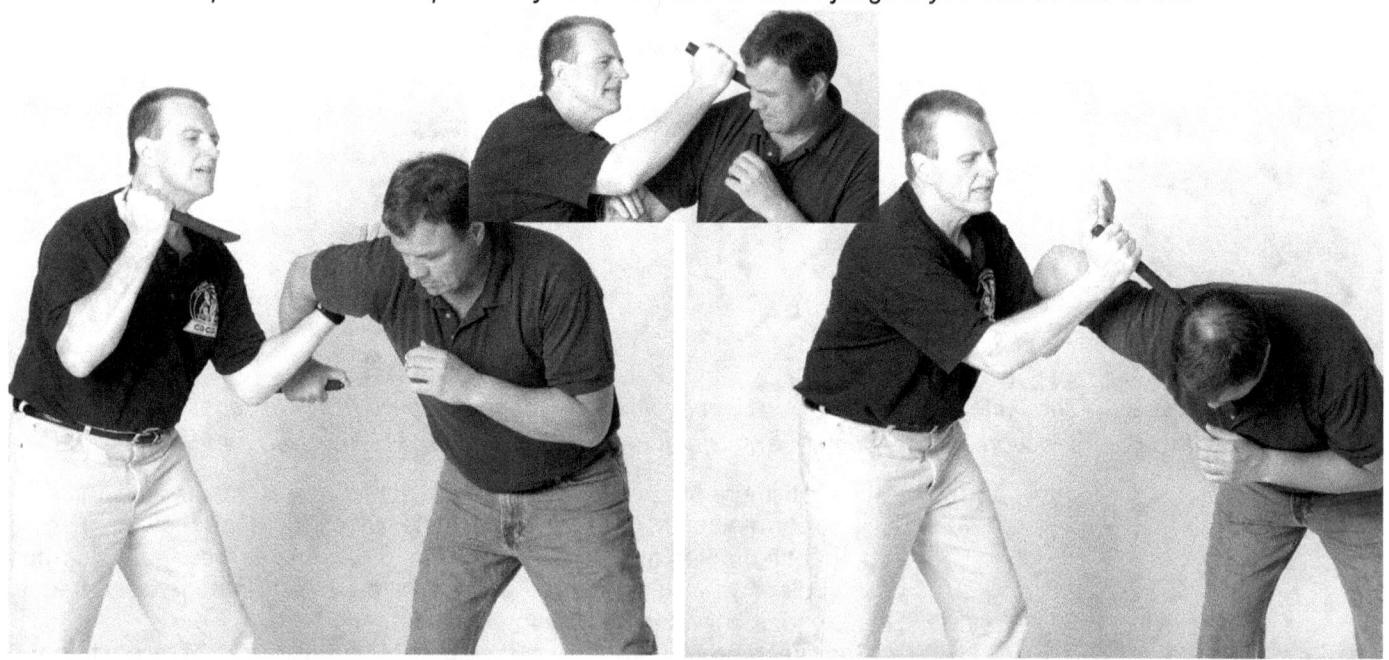

Hook the arm and stab the neck in the opening. This should diminish and confuse him, so you can finish the move.

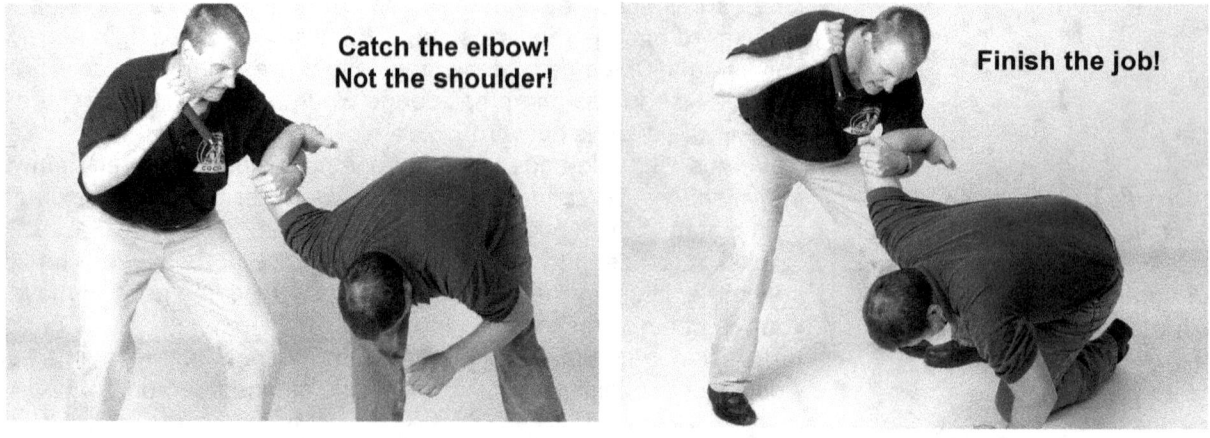

Catch the elbow! Not the shoulder!

Finish the job!

Advanced Windmill Drill Pattern Variations:

Early on in *Training Mission One*, I introduced the concept of two assault '"speed" attacks. The follow-thru attack and the hit-and-retract attack. These two attacks can be expressed in this Windmill Drill and are very important. In order to help me recall the practice, I have nicknamed them *Full Windmill* and *Retracted Windmill*. Understanding these two assault speeds are vital to preparing methods against a knifer.

The 10 and 2 Attack Drill: Full Windmill
This involves changing the angles up. The classic windmill comes in and down at your 2 o'clock angle. If you insert some 10 o'clock angles, you allow practice for common reality attacks. This is indicative of the fully committed angry attacker who drives his knife hard.

The 10 and 2 Attack Drill: Retracted Windmill
This involves changing the angles up. The classic windmill comes in and down at your 2 o'clock angle. If you insert some 10 o'clock angles you allow practice for common reality attacks. This is indicative of the "plucking" attacker who stabs and retracts. He retracts his knife fast and attacks fast. We really emphasize the reaction in, "In the Clutches" Module in *Training Mission Nine*.

*(These two versions will be studied again in the stick course (pommel attacks) and in **Training Mission Nine**, unarmed versus knife and stick.)*

A classic 2 o'clock attack.

A classic 10 o'clock attack.

The Science of the Windmill Attacker.
Here are the possibilities. It would be wise to practice with and against all of them in your knife classes. Follow-up grappling options will change with each factor.

Right hand 10 o'clock
Right hand 2 o'clock

Left hand 10 o'clock
Left hand 2 o'clock
Right versus left
Left versus left

Follow-Through Attacker
Hit and Retract Attacker

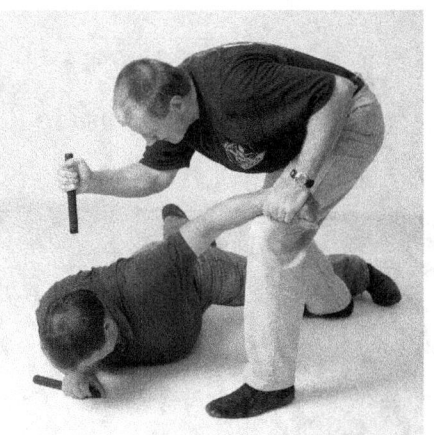

Grappling, finishing moves will change with each option.

Variations - The 10 O'clock Angle Attack

Since the mad-man, overhead attack may include ANY attack from above, we must see and prepare for these common assault angles. This includes the stabs bearing down from the attacker's 10 o'clock. In the windmill below, Jeff "Rawhide" Laun, follows the pattern then rears his knife over his head to deliver a stab from this new angle. I dodge and hook, then give him the same. The surprise angle inside the windmill pattern is good preparation practice.

The 10 o'clock attack. A good surprise angle to use and often aims right for the throat and sides of the neck.

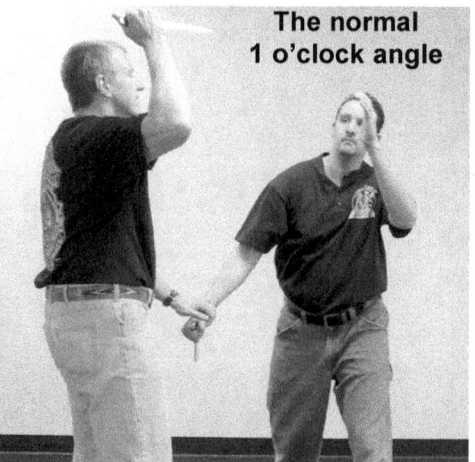

The normal 1 o'clock angle

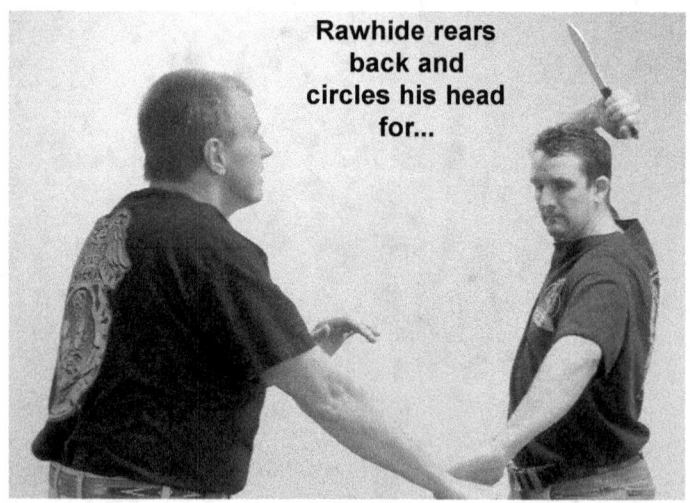

Rawhide rears back and circles his head for...

...the 10 o'clock angle stab.

You pass and dodge, and the pattern continues.

Variations - The 10 O'clock Crossover Hook

Since the mad-man, overhead attack may include ANY attack from above, we must see and prepare for these common assault angles. This includes catching the knife stab with a crossover hook with my 10 o'clock angle.

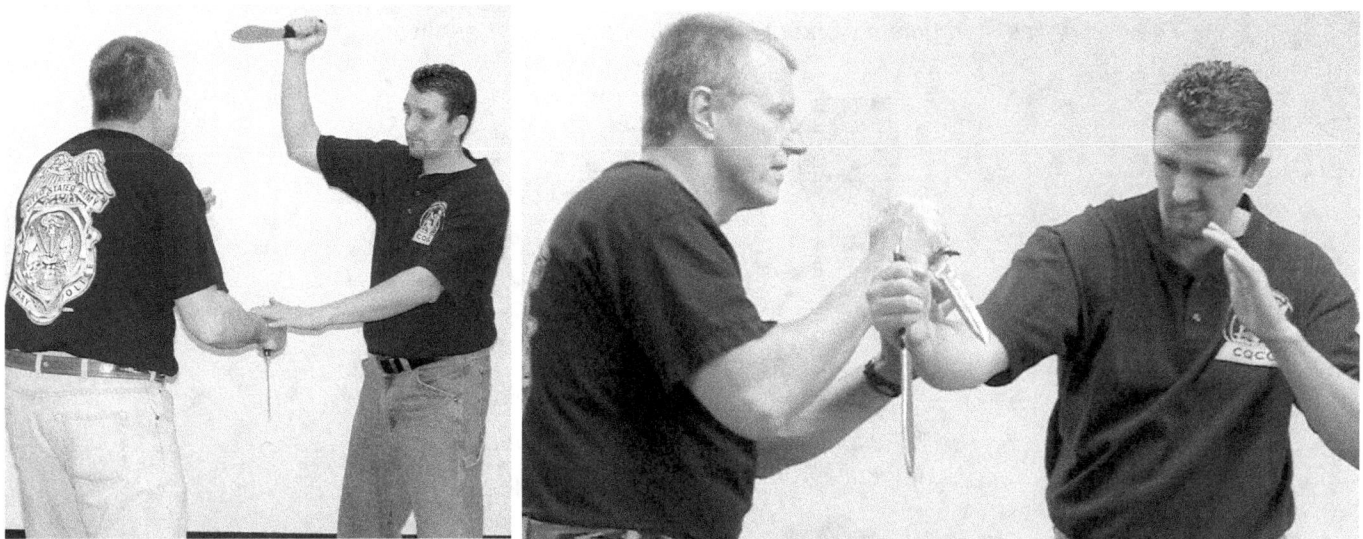

His stab comes down. You side step. You use your free hand to help pass, while your knife side crosses over and quickly hooks the downward stab. Pinch the blade upon the forearm/wrist.

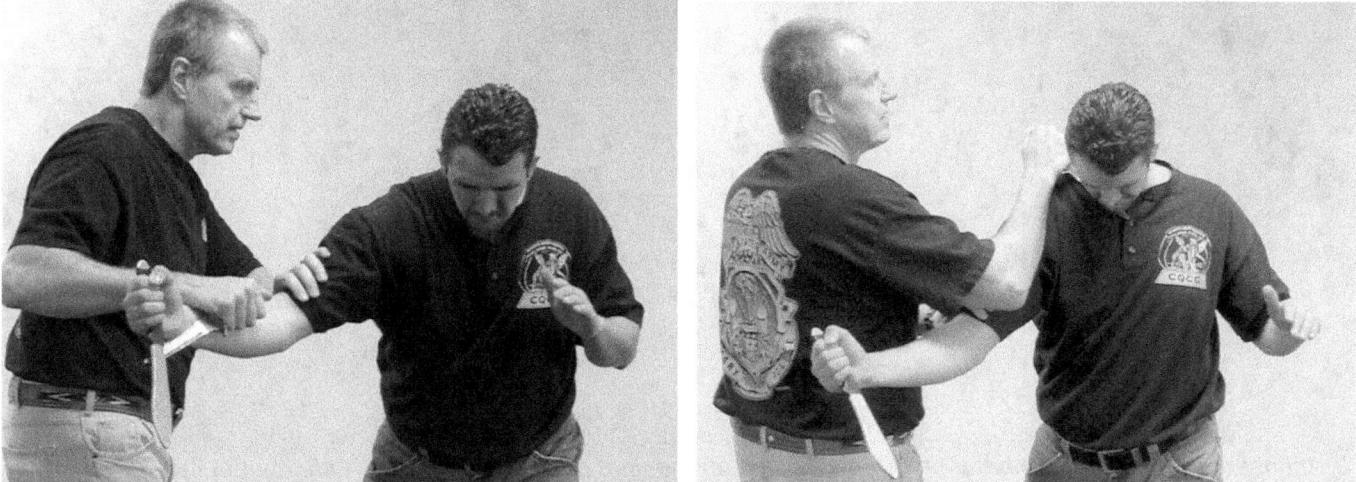

Your free hand grips his limb. Next, you engage in the typical, classical move - a throat attack. In your training you may let him try to stop this attack and work drills to counter this attempt.

Windmill to Chain-Saw

Imagine holding a chain-saw in front of you--the teeth rotating. This is another famous European and hence-Filipino skill drill. It takes but a few seconds to learn. The designated partner in the Windmill Drill brings the pommel to his solar plexus and thrusts the knife tip forward. The other partner opens his hand in a "c" clamp and stops the attack. The attack is quickly retracted to prevent a snare. This cycle continues. Eventually, one party executes an insert.

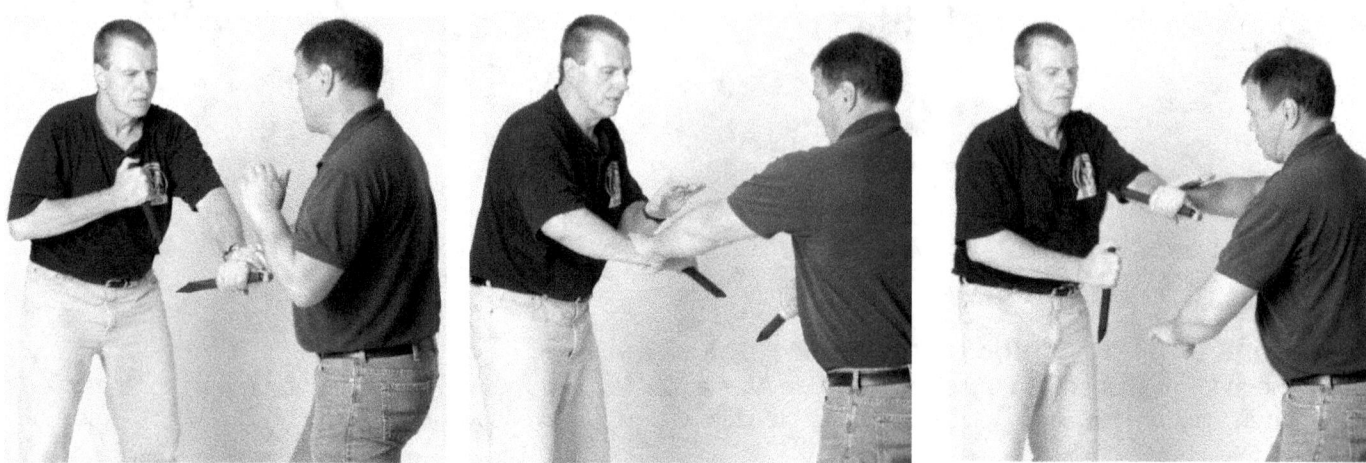

The center-line stab gets a workout in the Chainsaw Drill. Keep working these steps.

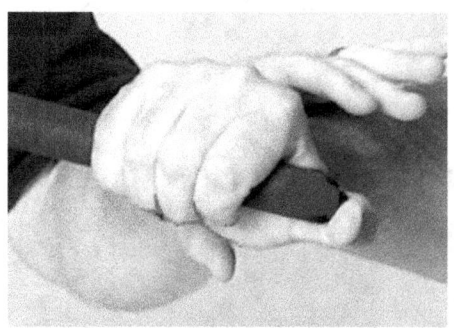

The "C" clamp catch with the thumb open, prevents the knife from slipping under a hand stop.

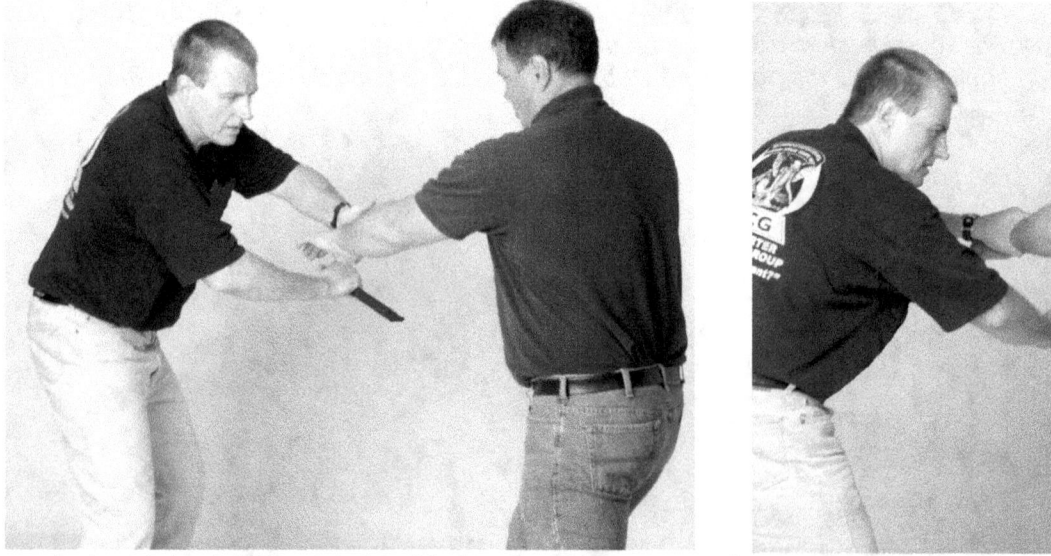

An insert. I catch Randy's hand before it solidifies on me, in a push pull motion. I yank it out of the way and deliver a stomach stab or two. From here any number of grappling techniques may be applied. By now, any student working up to this level should have grappling solutions.

CQCG

GUN/COUNTER-GUN COMBATIVES
LEVEL 5: Long Gun Disarming

"This is when many cops are killed," I said to myself as I stepped up to Harvey on his front lawn. "I think I can disarm him without shooting him. He's drunk and this is a domestic. Yep. This is where it can and usually happens."

Harvey clenched a shotgun, rip-roaring mad and again that full of whiskey. His real target was the woman from Dallas about half his age throwing her suitcase into the car Harvey had just bought her. She was leaving him in that very Camaro, and Harvey was primed to kill her, maybe himself and me too! But I had arrested him several times and knew him well enough that I might talk him out of the situation. The barrel and his steaming eyes were mostly on the girl.

"Harvey. Let it go. Give me that shotgun," I said, working myself closer. To which he cussed incoherently, barking words like, "get back in this house," to "I'll kill you." When I could see and feel that his whole physical attention was on her, not me, I jumped forward and grabbed the weapon with both hands. In a forward rowing motion I disarmed the gun from his grip. Another officer ran up from the lawn and snatched an arm. We cuffed him just as the girl slammed her door and raced off down Morse St. We just let her go, and I arrested Harvey for disorderly conduct, a typical charge for the times, circa 1978.

The rowing motion was a long gun disarm that my Drill Sergeant showed us when I was in Army Basic Training at Fort Polk, LA. But that was during the Vietnam training era when vets were teaching all kinds of CQC from both in and out of the manuals. I was later cajoled into showing some of the guys this movement at the police station and quickly realized that disarming rifles and shotguns was not taught to any of us at the police academy. Other prior military cops said they had not been taught a single rifle disarm. Nor was it taught in the Karate classes I had taken. Thank you, Drill Sgt. McCaskill, a wounded Vietnam vet with two plastic Dupont knees, who cared enough to pull a platoon aside and show us something he thought was important. In this spirit of education I undertook a scientific course of study in this problem. Here is what I have found.

Since the inception of the long gun, criminals and soldiers have killed, wounded, aimed at, threatened and kidnapped, escorted or otherwise controlled people with its ominous barrel. This project will study countering the rifle threat by scientifically identifying the probable confrontation positions and then problem-solving them. The long gun as defined here will be a rifle, shotgun, semi-auto and automatic firearm.

Looking back, I could have shot and killed Ol' Harvey that day with my revolver. No one would have argued the point, certainly not the frightened, onlooking neighbors that summoned us. Hell, he could have killed me! It's a call we have to make in the police business.

Harvey, minus the girl and the Camaro, lived to be a ripe old age and died naturally in his sleep, as I hope you will too after learning these tactics and strategies in case you need them. Oh...and good night, Sgt. McCaskill ...wherever you are.

Long Guns in Crime

According to research by United States police authorities, the use of rifles to assault and kill citizens and police officers is rising. Determined and calculating criminals have taken up the long gun. Many officers killed with rifles are shot when approaching an incident and while having close interaction with armed suspects.

The largest proportion of officers killed with rifles were shot while serving warrants or executing traffic stops-many of which were shot before they exited their patrol cars. Most shot with rifles are hit from the front with almost half of these shot in the head.

Of course, the business of military action surrounds the long gun. Citizens in many countries of the world live in fear of rifle-bearing, dictatorial, fascist and communist regimes. They dread invading armies with an eye for ethnic cleansing, rape, robbery, maiming and murder.

Physical Problem 3) Positioning
How is he standing and how is he holding the long gun? The enemy will present his long gun in six basic positions with numerous variations:

Position 1) In front

Position 2 and 3) The sides of you (right or left sides)

Position 4) Behind you

Position 5) Above you in some manner

Position 6) Below you in some manner

Physical Problem 4) Carry System
How is the enemy carrying his or her weapon? Is it merely held in their hands? Or worse, secured by a sling? There are three basic ways your opponent will be holding the weapon.

- Presented with hands
- Presented with sling
- Presented with secured lanyard/harness

Hand Carry
Criminals often use "civilian" guns like hunting rifles and other weaponry, firearms stolen in common burglaries. Civilian criminals will simply carry their weapons, foregoing slings and retention holsters, making quick disarms more possible.

Sling Carry
I have conducted an intensive study for years of military history, studying photographs of highly trained and sophisticated international troops, as well as untrained rebels. In taking note of some thousands of photos or armed military personnel, about half engage in the use of their slings, and half ignore the sling and it dangles below the weapon. To focus more on the subject, many of these photos involved prisoner acquisitions and escorts. A slung weapon wrapped around a body part of the enemy poses an obstacle to the disarming of the weapon. Military personnel have slings. The primary purpose of slings is to cart the weapon in both "stand-down" and offensive manners.

Later, it was discovered that slings could be used to support marksmanship efforts. Also, the sling is grabbed at the barrel connection for quiet and safe carry during a low crawl.

The slings allow for:

- across chest carries

- under armpit carries

- over shoulder carries

- across armpit and shoulder carries

- clipped carry: In the last decade, vests and support gear such as lanyards have become very popular. These pose another problem during the disarming process.

- ground crawling

Long Gun Attacks

A trained enemy holding a long gun on you can attack you in many ways. He may:

1) Threaten and control you

2) Shoot you
- with a full range of calibers

3) Non-ballistic attacks
- impacts strikes from barrel to stock
- related bayonet stabs and slashes
- pushes, pulls and chokes

Distance 3 Ranges of Weapon

How far away is the enemy? Learn to gauge the distance to the barrel, to the trigger finger, to the stock and to the body.

Range 1) In Physical Contact
Range 2) Lunge and Reach Distance
Range 3) "Sniper" Distance (verbal disarming possible?)

Status of Weapon after Disarm

Once you attain the weapon, do not trust it to work. It may be unloaded. It may be a replica. It may be knocked out of battery in the struggle and with the great variety of long guns, you may not be qualified to get it back into operation. Plus the gunfire may draw attention to your success and bring his comrades. You may have to contain him with makeshift ligatures or even kill him if the situation justifies. Once secure, and if time permits, search him and confiscate all his weaponry and support equipment.

Status 1) Is it a replica?
Status 2) Is it unloaded?
Status 3) Is it out of battery from struggle?

Resort to your own weapon if possible. Use disarmed weapon for impacts.

Main and Common Disarm Solutions
More often than not, these are the steps of a successful long gun disarm

> Solution 1) Always clear the barrel from you and if possible-others.
>
> Solution 2) Always bash the holder, neck and head first if possible. Take the weapon from the stunned or unconscious man. You must review and hone all practical unarmed combative skills.

It is Time to Strike?
Many victims have escaped while being escorted to questioning, eating, restrooms or sleeping quarters. Many have surprised a tired or untrained guard. Many have waited until a guard was left alone. Many have known they were about to be executed and have decided to die fighting - and they won and lived.

So where the enemy stands, how he looks, what side of the body the gun is held and identifying their carry method are the main observations you must make prior to taking physical solutions in this worst case scenario.

Explosion
You are holding on to his gun, and the fight is on! What if he pulls the trigger? The explosion that occurs inside a weapon may be a problem. Obviously there's also a tremendous sound, which could hurt your ears. There is indeed a detonation in the weapon every time it goes off. Some experts claim 40,000 to 60,000 pounds of pressure blows in the barrel. There is heat that comes from an open cylinder possibly or any ports. Or if a gun has been run for a long time, like in some really prolonged combat situations, just grabbing the weapon my be a hot experience.

You may experience some hearing loss and possible burns on your hand in the process of disarming a weapon that is fired. Don't stop and don't let this interrupt the process. Don't jerk your hand back off the gun (which is statistically what happens). If you've got the gun, hang in there and keep working. It's better to have a slight cut or burn on your hand than it is to let him step back and blow your brains out.

> Sound! Blast!
>
> Burn from any open cylinders or ports (or HOT gun).
>
> Hearing loss.
>
> Choices? Minimization of wounds theory.

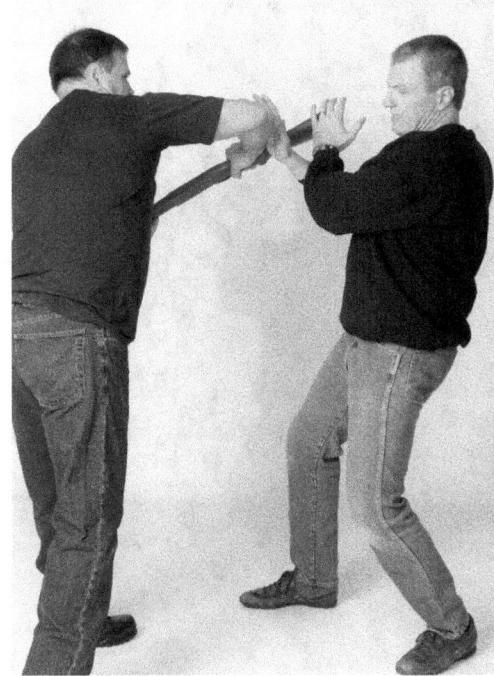

Keep Your Surrendering Hands Moving

When confronted with an armed, threatening assailant. Give him the hands-up and palms-out surrendering hands. Do not hold them still! The sudden attack movement with your hands may cause a startle reaction and from this startle he may well shoot you, even if he didn't plan to! Instead, keep your surrendering hands in motion, nervously up and down, side-to-side, to disguise any attack. Versus a long gun, the position of your hands can really help the acquisition of the gun.

Versus a long gun, the position of your hands can really help the acquisition of the gun.

How the Enemy will Threaten and/or Attack You with a Long Gun

Study these positions. They include pulling the weapon up or pulling it into position from a sling carry, striking with a long gun (the 15 angles of SDMS attack), presenting the weapon to threaten, and actually shooting the weapon.

> **Attack Method 1)** Drawing the gun into action from non-ready and ready positions.
>
> **Attack Method 2)** Striking with the weapon (also stabbing with a bayonet attached).
>
> **Attack Method 3)** Threat positions.
>
> **Attack Method 4)** Firing the weapon.

Attack Method 1) Drawing the Gun.
Weapons are carried across the chest, between the arm and the torso, and across the back. Long guns are also snatched from nearby lunge and reach positions. Please refer and review back to Gun Level 1 for the section on *Dismounting* - a term used for taking a sling weapon from a carry position and raising it up into use. Start your study there and return to the next method.

Please refer back to TM1 book and DVD set to see the sling carries and long gun presentations.

Attack Method 2) Striking with the weapon (also stabbing with a bayonet attached).
This is the close quarter impact weapon approach to utilizing the long gun as a striking implement. As rifles evolve into high-technology platforms of scopes, laser sights, flash lights and other scientific gear, it becomes less likely one will use a barrel smash or a butt stroke as with an older M-1 style rifle.

Learn the 15 main ways a person will impact strike you with a long gun.

Attack Method 3) Threat positions.
You will have to take action while being held *under the gun* by threat of a shooter controlling or guarding you.

Attack Method 4) Firing the weapon.
You will have to take action at the moment the decision is made to shoot you and/or while being fired upon.

Learn the common ways people hold this weapon to shoot you.

Potential strike positions, as organized via the DMS 15 Strike System

1) High right strike 2) High left strike 3) Mid right strike 4) Mid left strike 5) Horizontal strike

6) High right stab 7) High left stab 8) Low left strike 9) Low right strike 10) High right hook

11) High left hook 12) Downward strike 13) Stab 14) Upward butt stroke 15) Upward bayonet slice

Hip shot    Low shoulder shot    High shoulder shot

In the tactics and scenarios ahead, we will battle all these individual attacks.

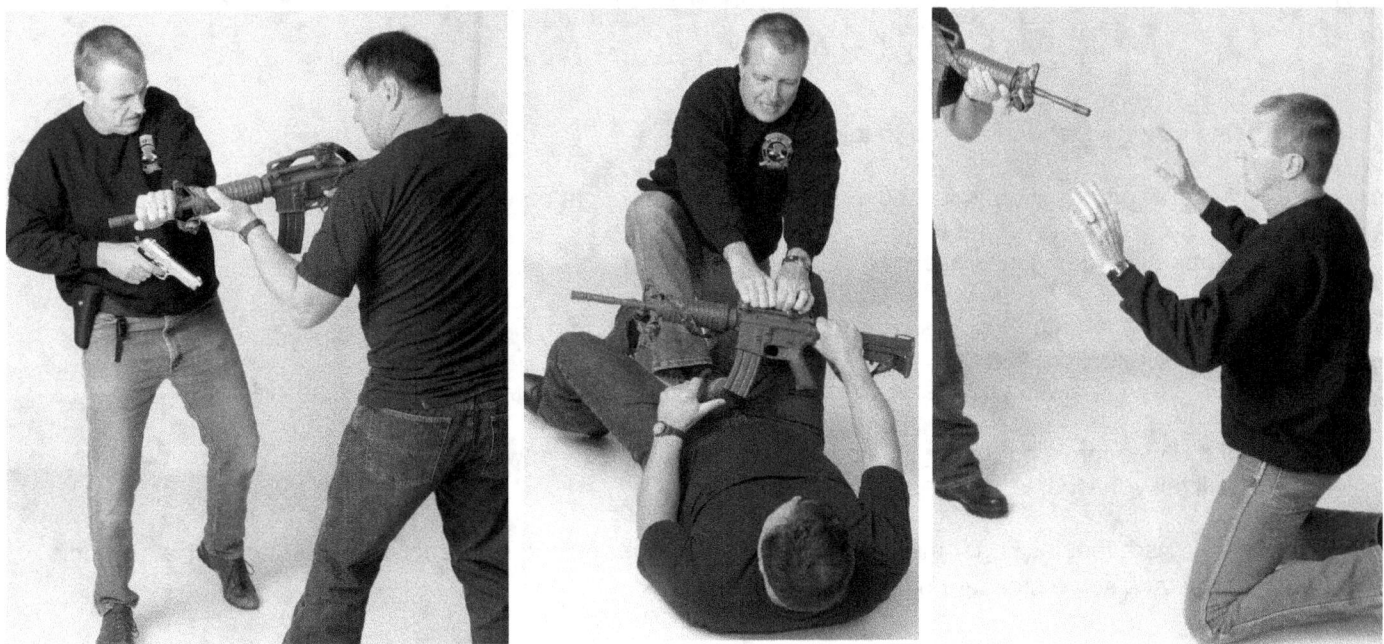

Long Gun Disarm Scenario 1: Escape the Rear Escort

You are being escorted, and you need to take action.

1) Note whether the enemy is holding the weapon in a right or left handed grip.

2) Note if the long gun is attached by a sling or lanyard.

3) Note how long the stock is. This matters when wrestling over the weapon.

4) If the gun is low in your back, keep your surrender hands low. Gun high? Hands high.

5) If the weapon is not giving you consistent pressure, slow down, and let the enemy poke you. This will tell you where the weapon is.

6) Be aware that his finger may be on the trigger. If you push the weapon back during your attack, this may take some pressure off the trigger.

He pushes. Clearing your back of the barrel, you arm sweep to his outside. Start seizing the weapon as you...

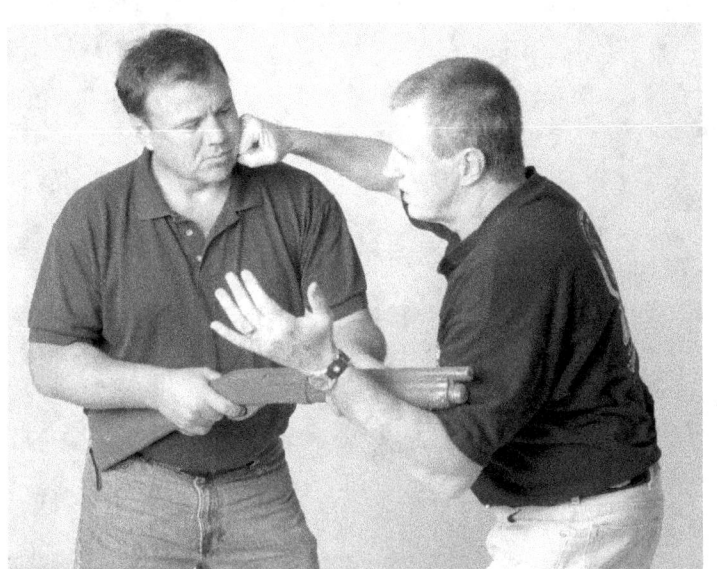 

...punch his throat. Batter his arms free of the weapon.

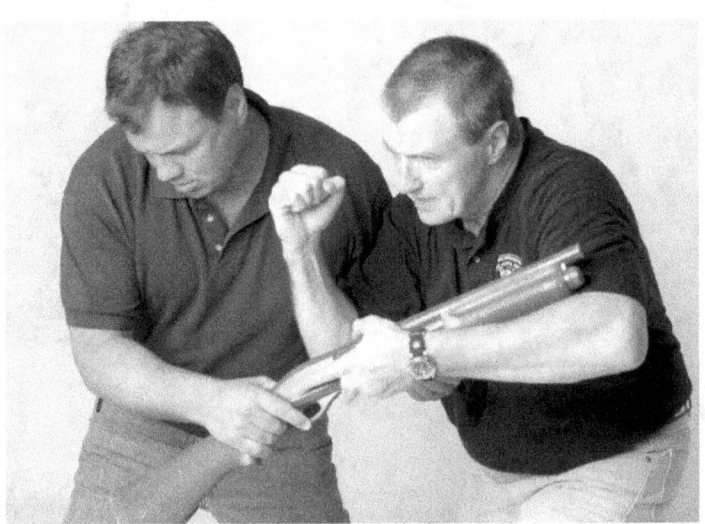 

Use the stock or the barrel to batter the enemy down.

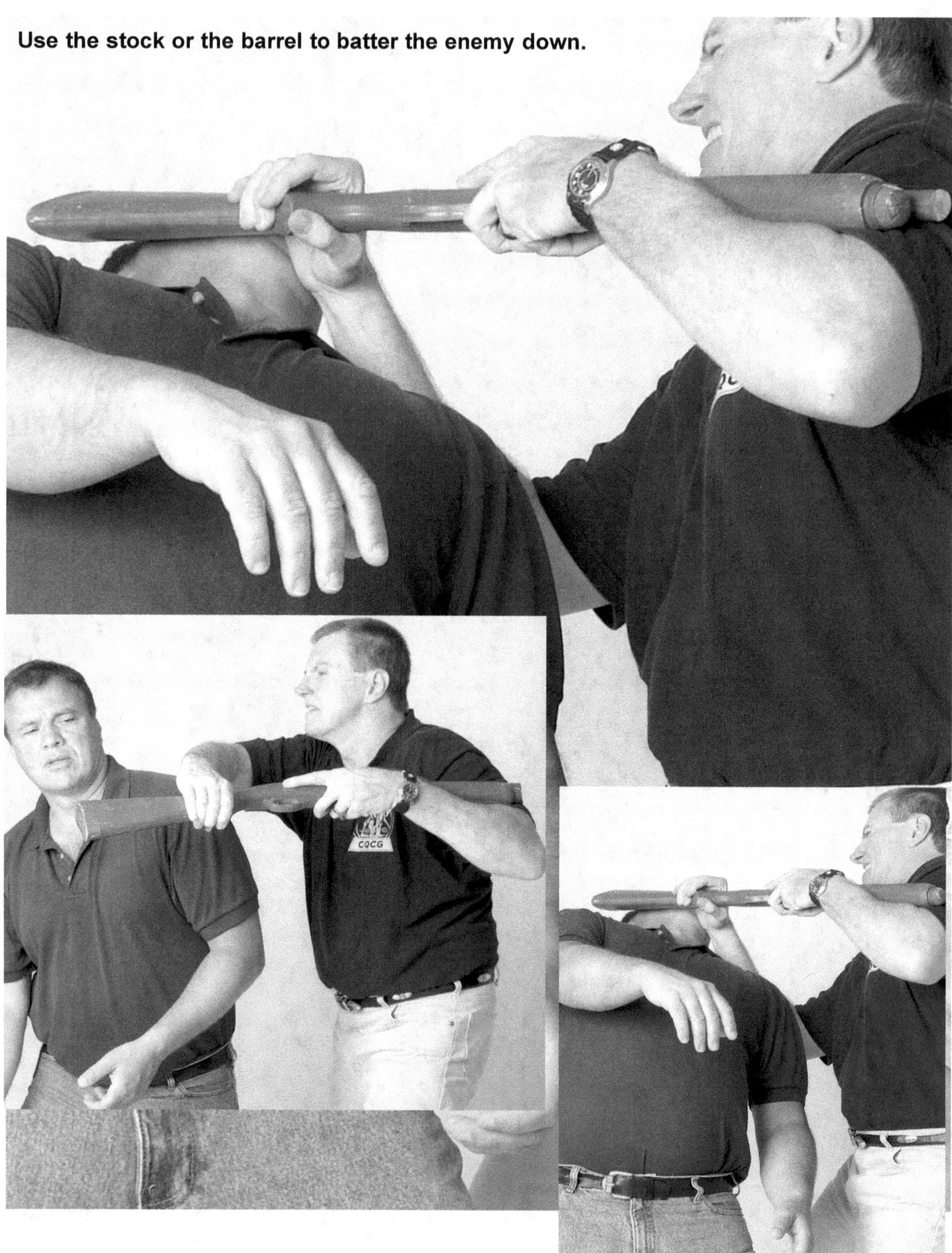

Long Gun Disarm Scenario 2 : The Sling Throw

If your enemy is wearing a sling or his weapon is attached to him by way of a lanyard clipped to a tactical vest, you will have a problem disarming him with many of the conventional methods shown by many systems.

You must use this connection as a lever to throw the stunned enemy.

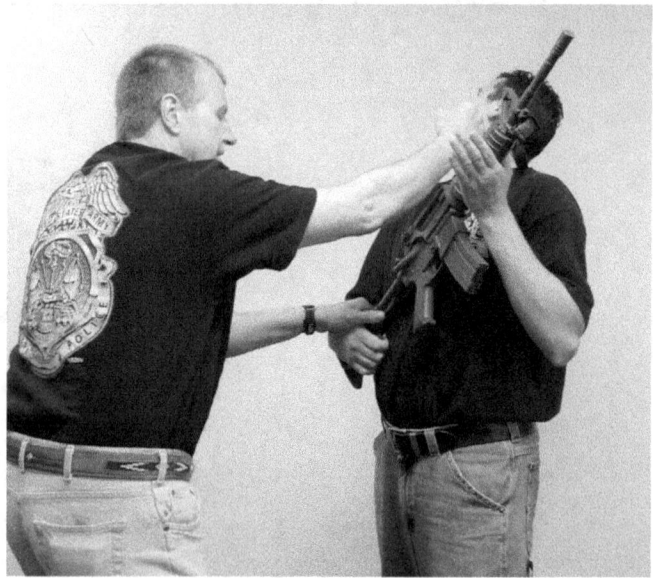

Get close enough to strike. Stun the enemy sufficiently. Grasp the weapon and use the attached sling or lanyard to throw him down. Continue to strike. Think of using the attached weapon as a lever to throw the man down.

You may have to use conventional grappling, foot positioning and/or reaps to assist in the takedown. Next, remove the sling or lanyard and weapon from the incapacitated man. Check for weapon operability and ammo.

Long Gun Disarm Scenario 3: Hammer Fist the Guard

These steps are for times when you can approach and question a guard who is not alert. He may have pulled a long shift, or guarded your group for many days.

You may have practiced approaching him several times with innocent questions, then retreated. Your questioning presence before him may become somewhat routine.

You strike with double hammer fists on the carotid arteries, which alone could knock him cold. From this point on you batter him until you disarm him and win.

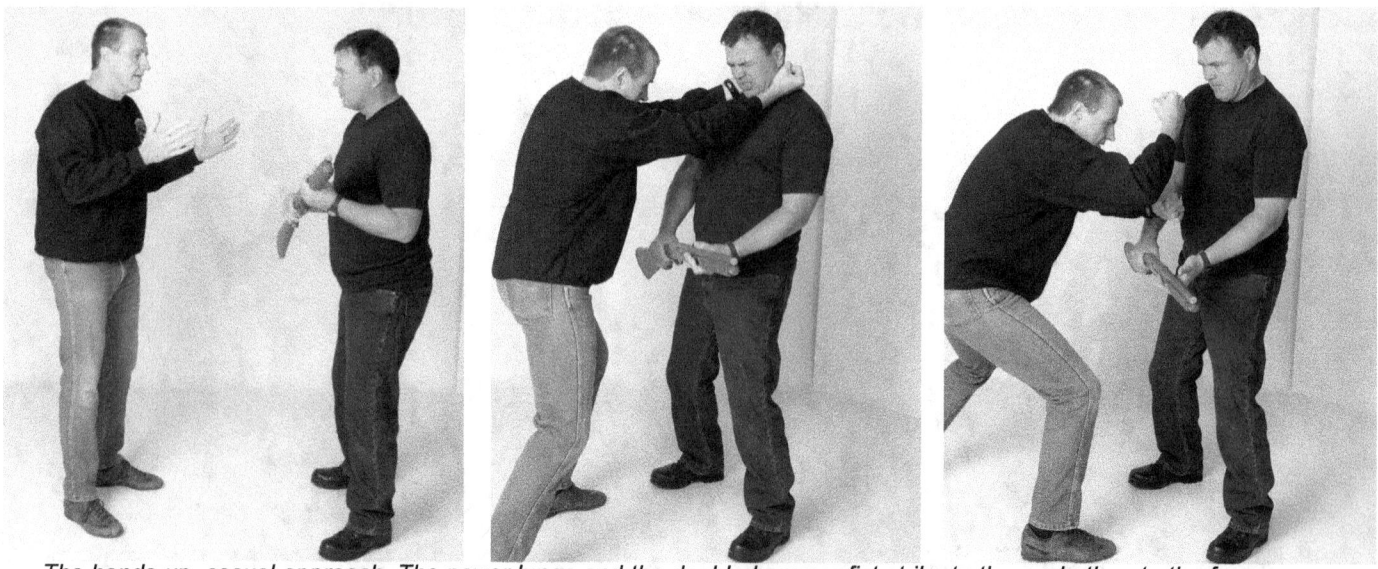

The hands-up, casual approach. The power lunge and the double hammer fist strike to the neck, then to the forearms.

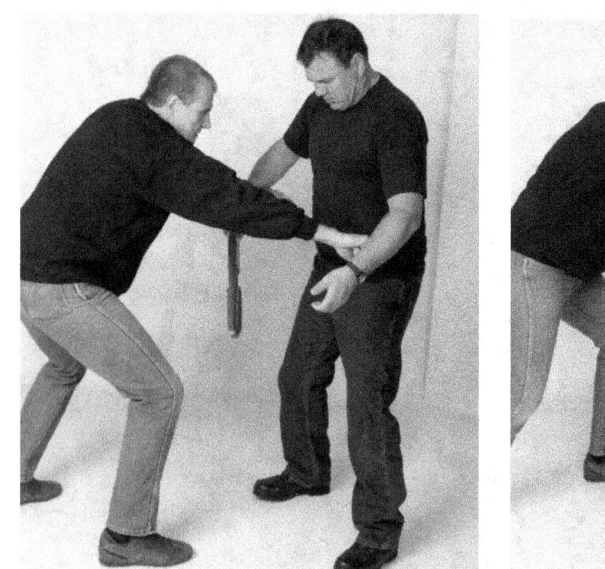

Bang the arms aside. Usually at least one hand hangs onto the weapon. Snake and keep.

You have a moral, legal, ethical and sometimes even a political obligation to follow the Use of Force Rules.

Long Gun Disarm Scenario 4: The Big Duck Under

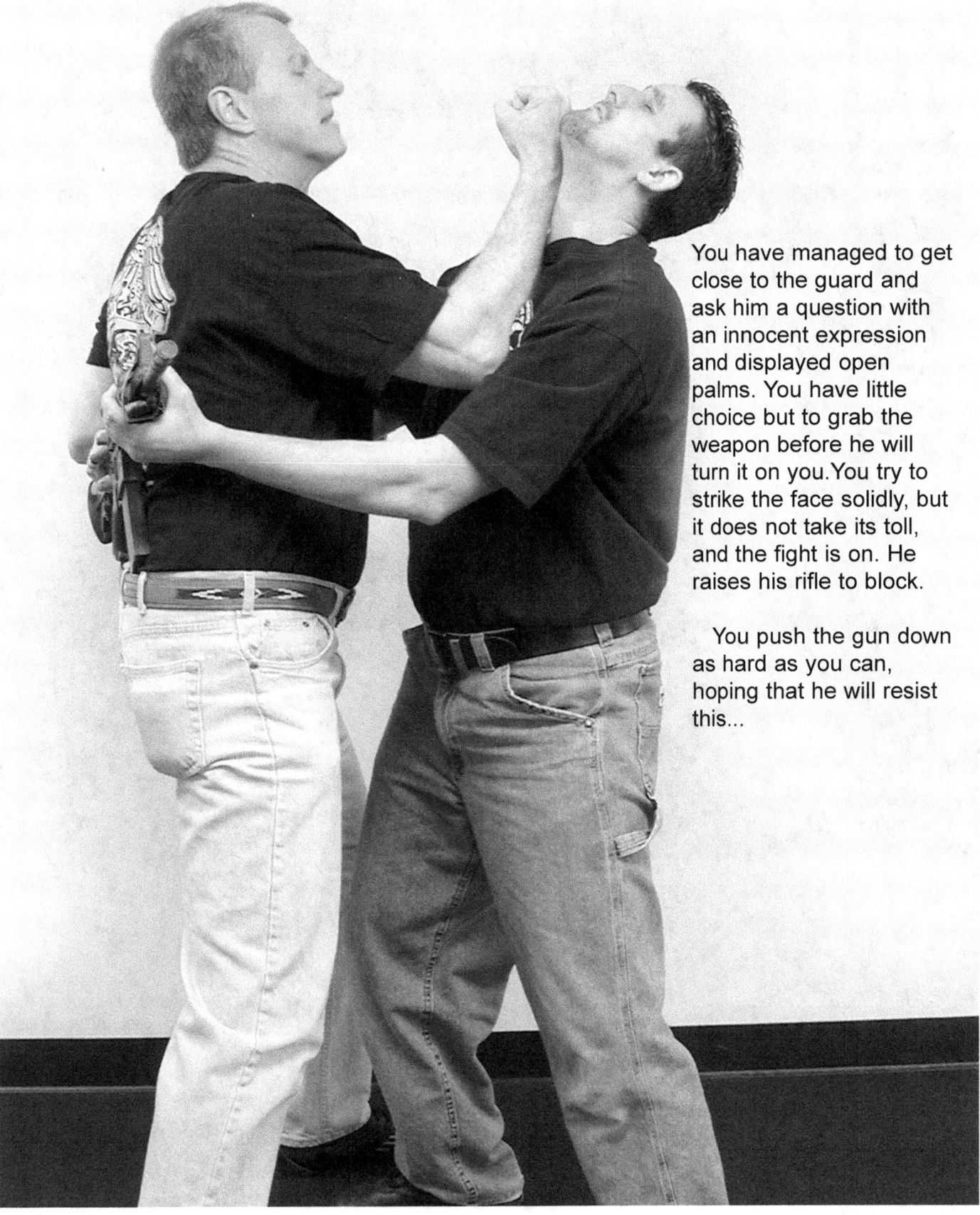

You have managed to get close to the guard and ask him a question with an innocent expression and displayed open palms. You have little choice but to grab the weapon before he will turn it on you. You try to strike the face solidly, but it does not take its toll, and the fight is on. He raises his rifle to block.

You push the gun down as hard as you can, hoping that he will resist this...

You have managed to get close to the guard and ask him a question with an innocent expression and displayed open palms. You have little choice but to grab the weapon before he will turn it on you. You try to strike the face solidly, but it does not take its toll, and the fight is on. He raises his rifle to block.

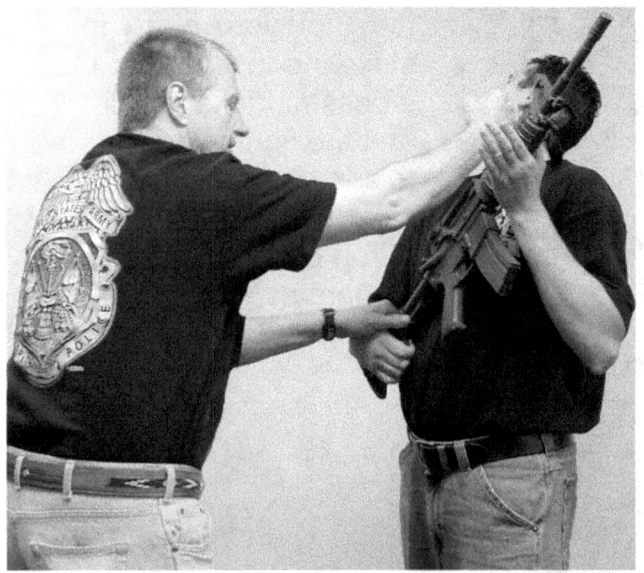
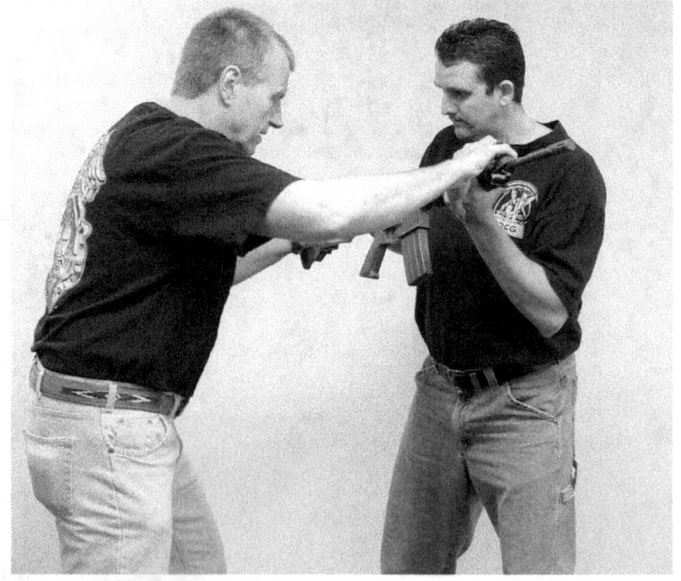

You push the gun down as hard as you can, hoping that he will resist this...

...by pushing his gun in the opposite direction. You will join this upward force.

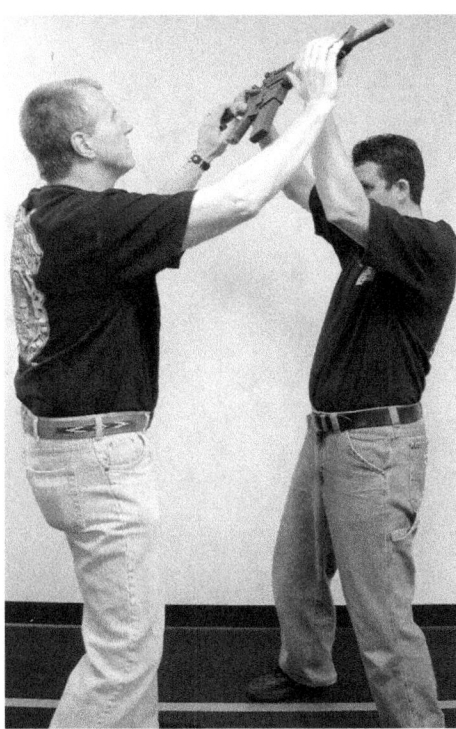  

You push up powerfully with his force and, as soon as possible, skip in under the weapon-fist first. Start beating viciously away at the enemy. I believe that the guard will be stunned and also somewhat committed into holding onto his rifle for as long as possible. It is common sense psychology to assume this and a good bet he will. Remember to beware of the sharp corners and ends on the weapon. Once he is stunned, make an effort to remove the weapon from his grip. If he has a sling, you may not choose this method, and may instead follow the "with sling" scenario in this chapter.

 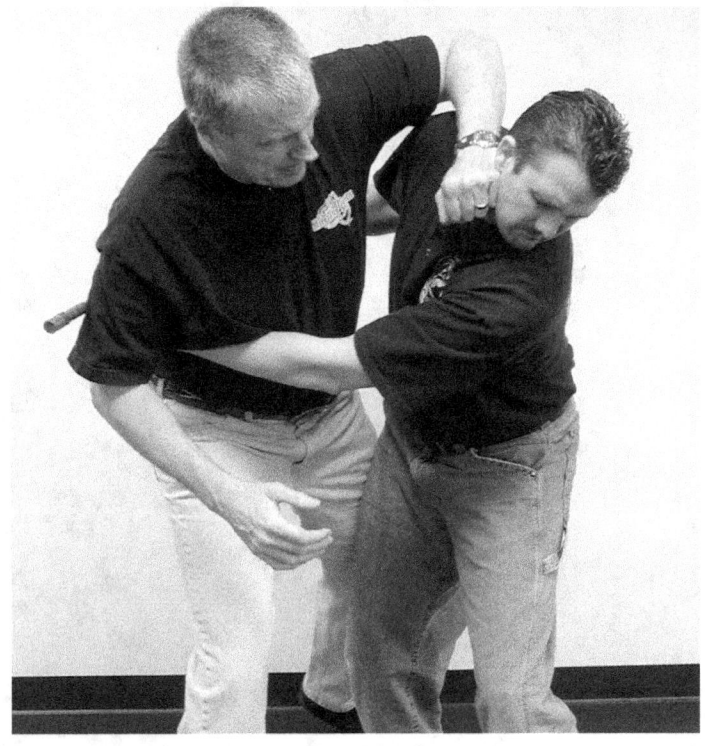

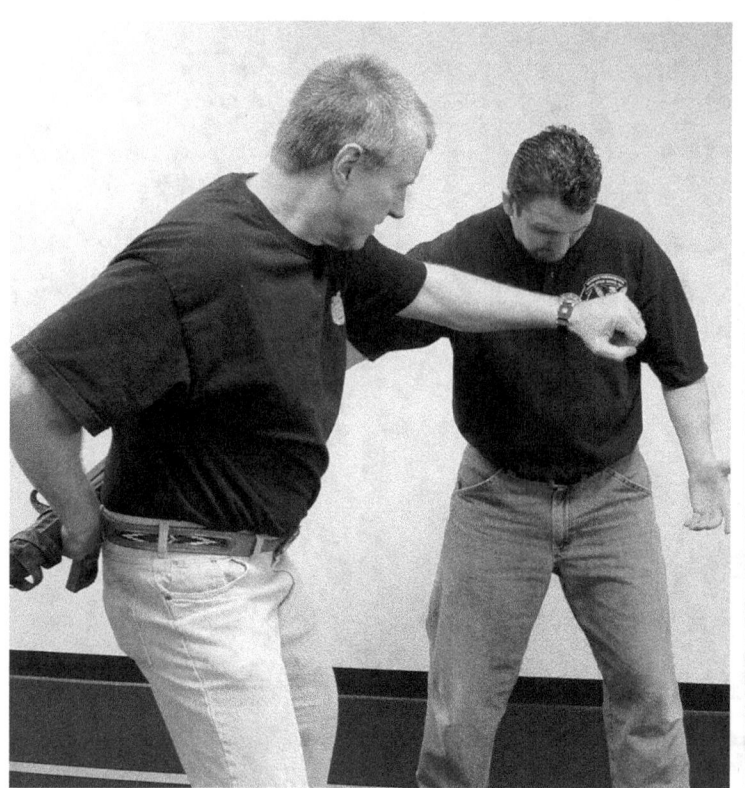

Long Gun Disarm Scenario 5: The Power Push and Roll-over

There was a four-hand struggle over this weapon and:

1) He made a mad dash at you, giving you great force.

2) Or, you may have wisely pushed hard. He returned with the same energy. You pulled with this energy.

3) Or, he may have charged so hard you started falling over backward.

Try to minimize the impact on your back. Learn this with practice. Pull and roll quickly. This usually pitches the opponent right over you, and you usually wind up as the one holding the weapon.

You plant one, or better, two feet on each side of his pelvis. Maintain your tight grip on the weapon. Keep your arms strong as in a bench press as he goes over and falls or drops back.

Page 175 - W. Hock Hochheim's Training Mission Five

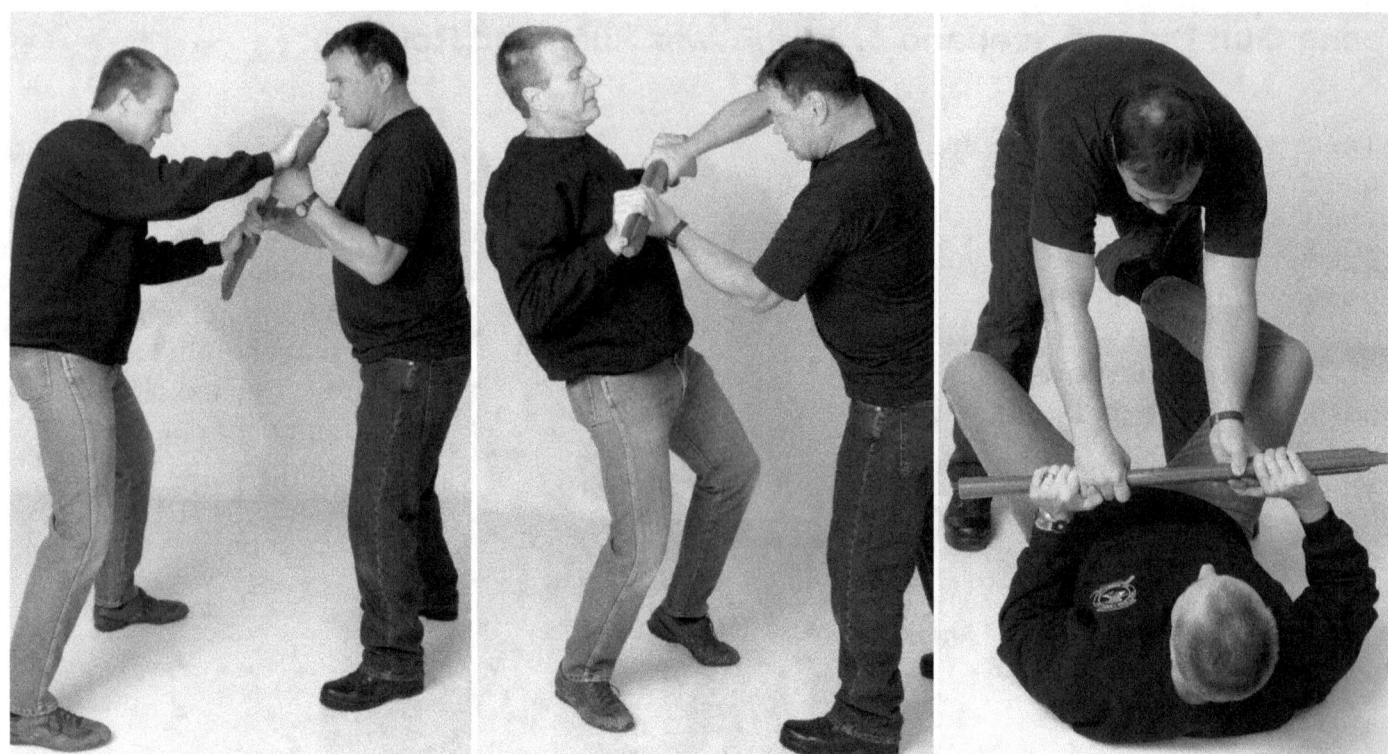

There was a four-hand struggle over this weapon, and you are falling. Try to minimize the impact on your back. Learn this with practice. You plant one, or better, two feet on his pelvis. Pull and roll quickly.

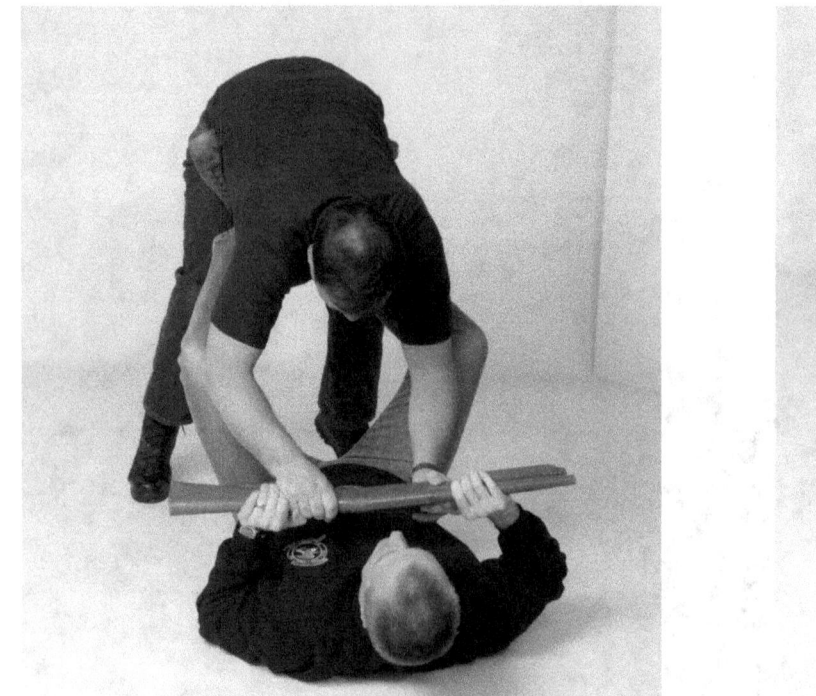
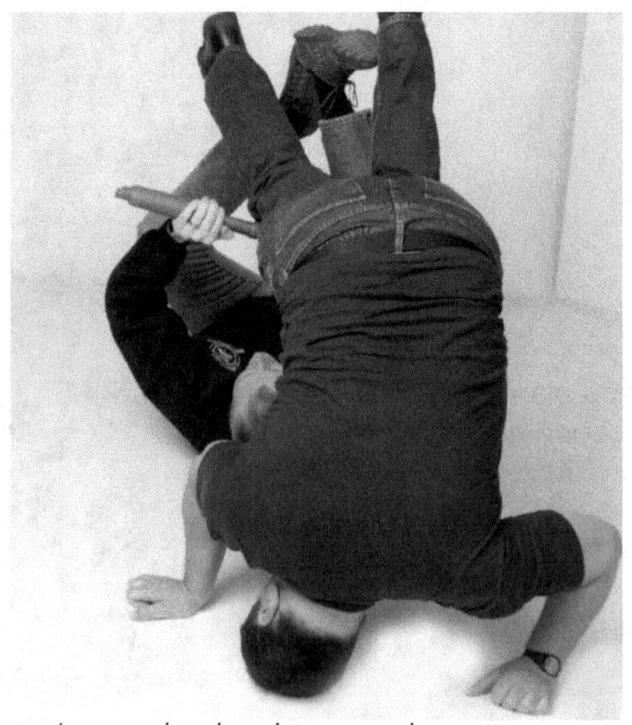

Maintain your tight grip on the weapon. Keep your arms strong as in a bench press as he goes over and falls or drops back. This usually pitches the opponent right over you, and you usually wind up as the one holding the weapon. Get right up. Take the smartest action.

Long Gun Scenario Disarm 6: Defeat the Bayonet Stab

The enemy stabs at you with his rifle and bayonet. Use all your athletic footwork skills to get out of the way and your arms to pass the weapon. The delivery of a bayonet stab, since it is attached to a long and sometimes heavy rifle, makes the attack cumbersome. It is often hard to work such long and heavy gear against an effective counter attack. As soon as possible, you blind the enemy with an eye attack or stun them with a serious throat or face attack. Then use a series of batterings to get that weapon out of his hands. The military trains this bayonet stab in conjunction with a powerful, charging run. Expect some vicious energy.

He is going for it. You must use all your athletic prowess to dodge the bayonet. Use your arms to help the effort by passing the weapon on the barrel. You must practice this evasion extensively as a separate drill.

The bayonet pushed aside? Charge.

At this point you must open a barrage of attacks on his face and neck until he is deeply stunned. Keep the bayonetted rifle tight between you.

When he is stunned? You pull the weapon away from him.

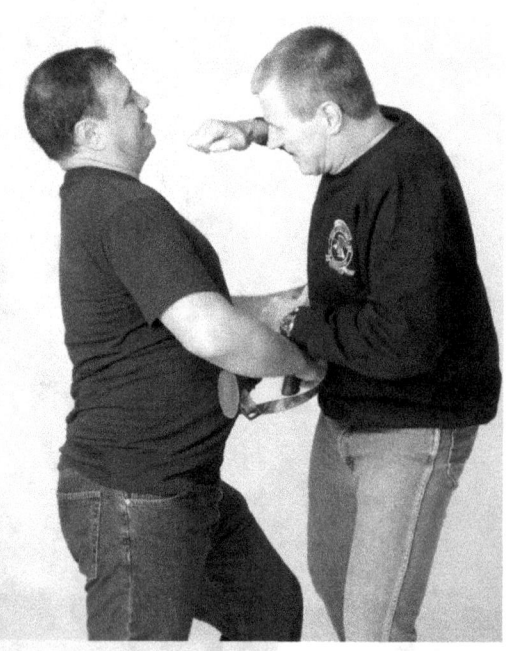
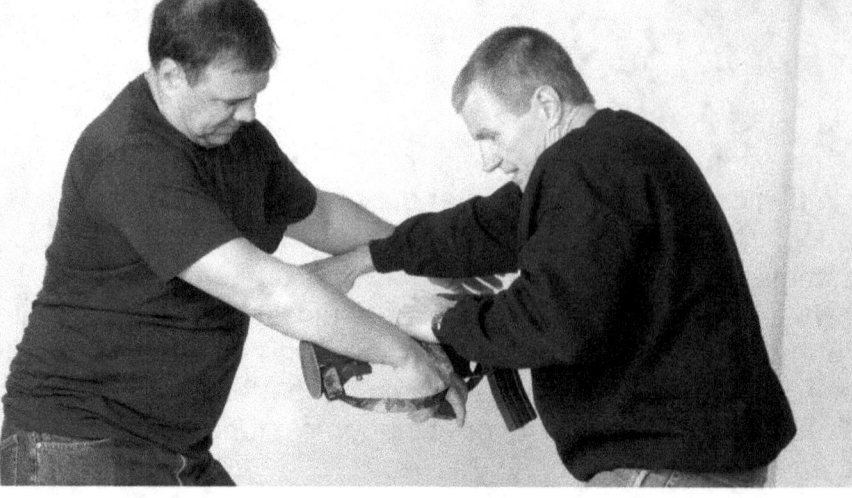

Page 178 - W. Hock Hochheim's Training Mission Five

Long Gun Disarm Scenario 7: Countering a Barrel Thrust

Follow the same strategies as before in the bayonet stab situation. Clear the barrel. Try to secure the barrel and then batter the enemy.

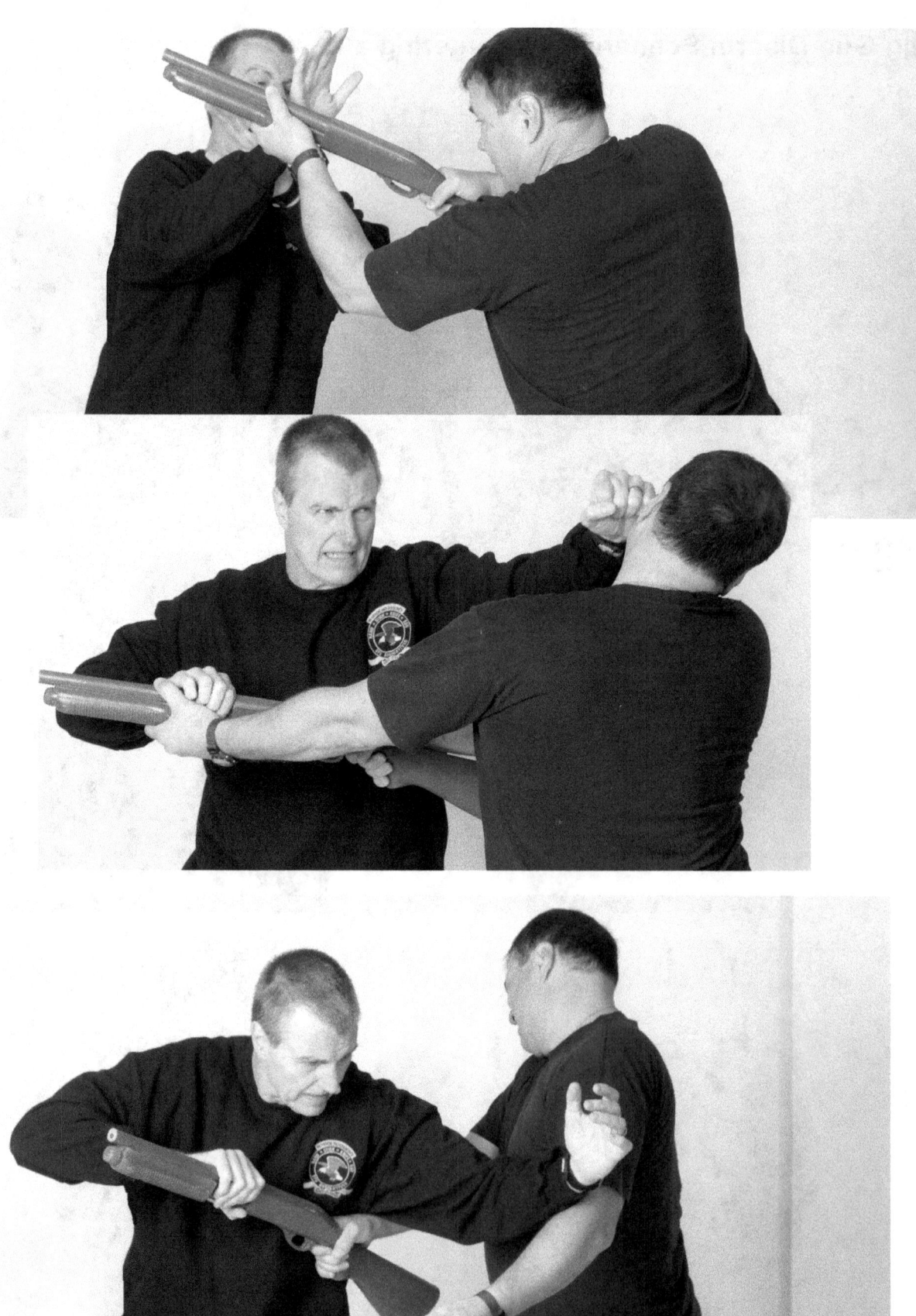

Long Gun Disarm Scenario 8: Escape the Choker

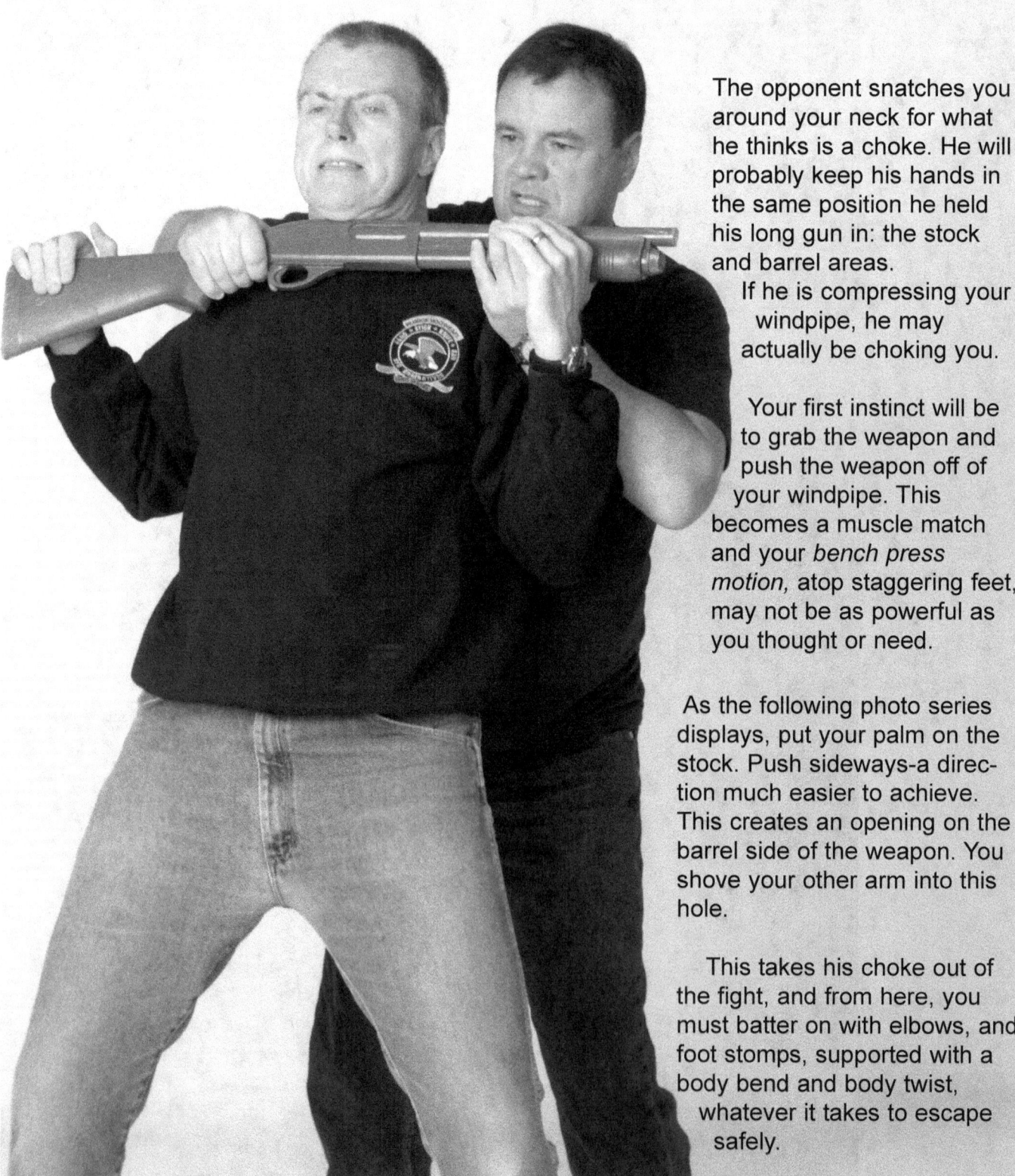

The opponent snatches you around your neck for what he thinks is a choke. He will probably keep his hands in the same position he held his long gun in: the stock and barrel areas.

If he is compressing your windpipe, he may actually be choking you.

Your first instinct will be to grab the weapon and push the weapon off of your windpipe. This becomes a muscle match and your *bench press motion,* atop staggering feet, may not be as powerful as you thought or need.

As the following photo series displays, put your palm on the stock. Push sideways-a direction much easier to achieve. This creates an opening on the barrel side of the weapon. You shove your other arm into this hole.

This takes his choke out of the fight, and from here, you must batter on with elbows, and foot stomps, supported with a body bend and body twist, whatever it takes to escape safely.

The catch!

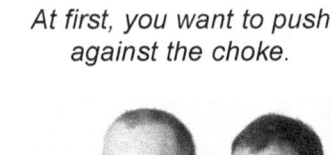

At first, you want to push against the choke.

But, you must push sideways on the stock. Not the barrel. He could shoot! Get an opening on the barrel side.

Ram a fist and arm into the new opening.

Push sideways-a direction much easier to achieve. This creates an opening on the barrel side of the weapon. You shove your other arm into this hole. This takes his choke out of the fight and, from here, you must batter on with elbows, and foot stomps, supported with a body bend and body twist-whatever it takes to escape safely.

Long Gun Disarm Scenario 9: Counter the Execution

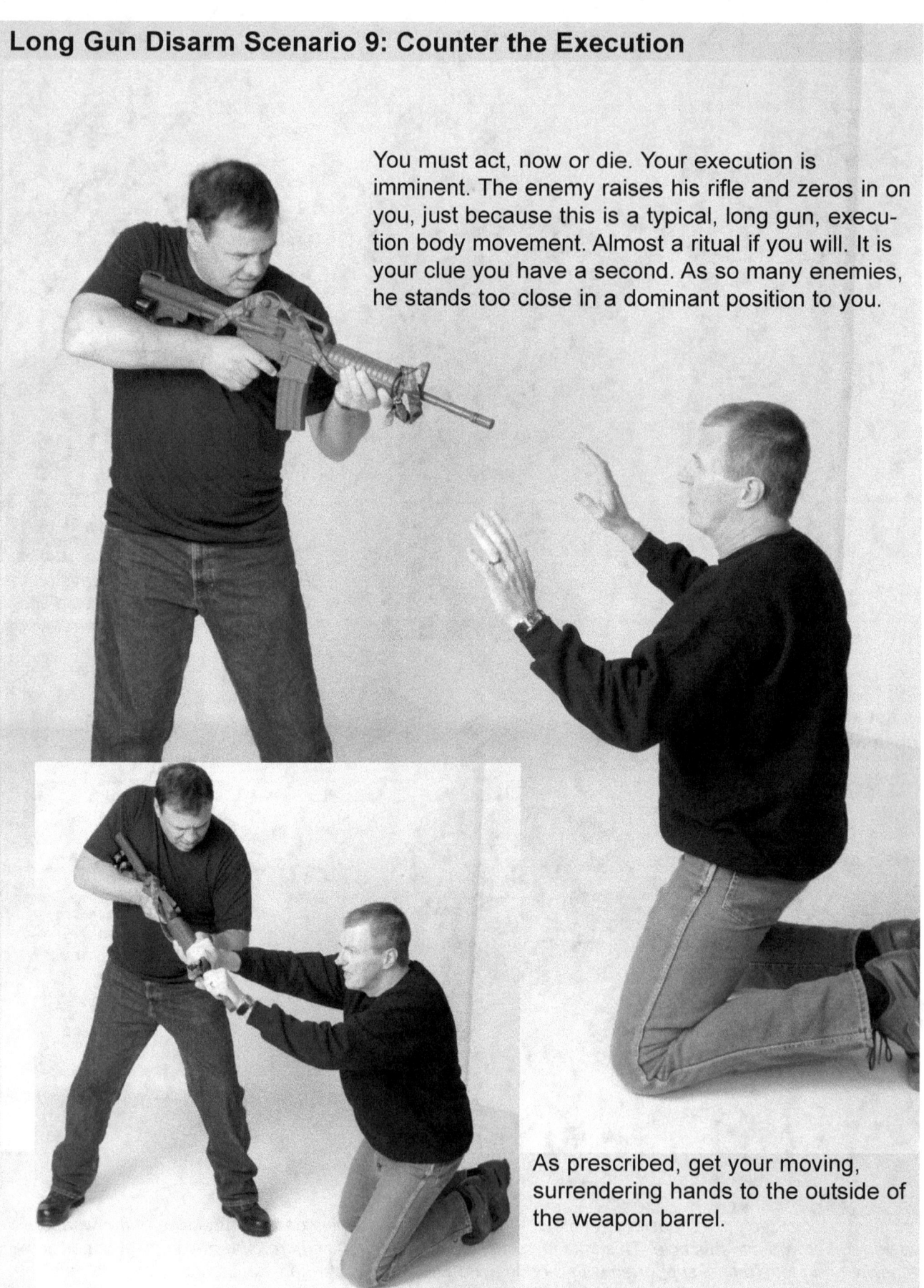

You must act, now or die. Your execution is imminent. The enemy raises his rifle and zeros in on you, just because this is a typical, long gun, execution body movement. Almost a ritual if you will. It is your clue you have a second. As so many enemies, he stands too close in a dominant position to you.

As prescribed, get your moving, surrendering hands to the outside of the weapon barrel.

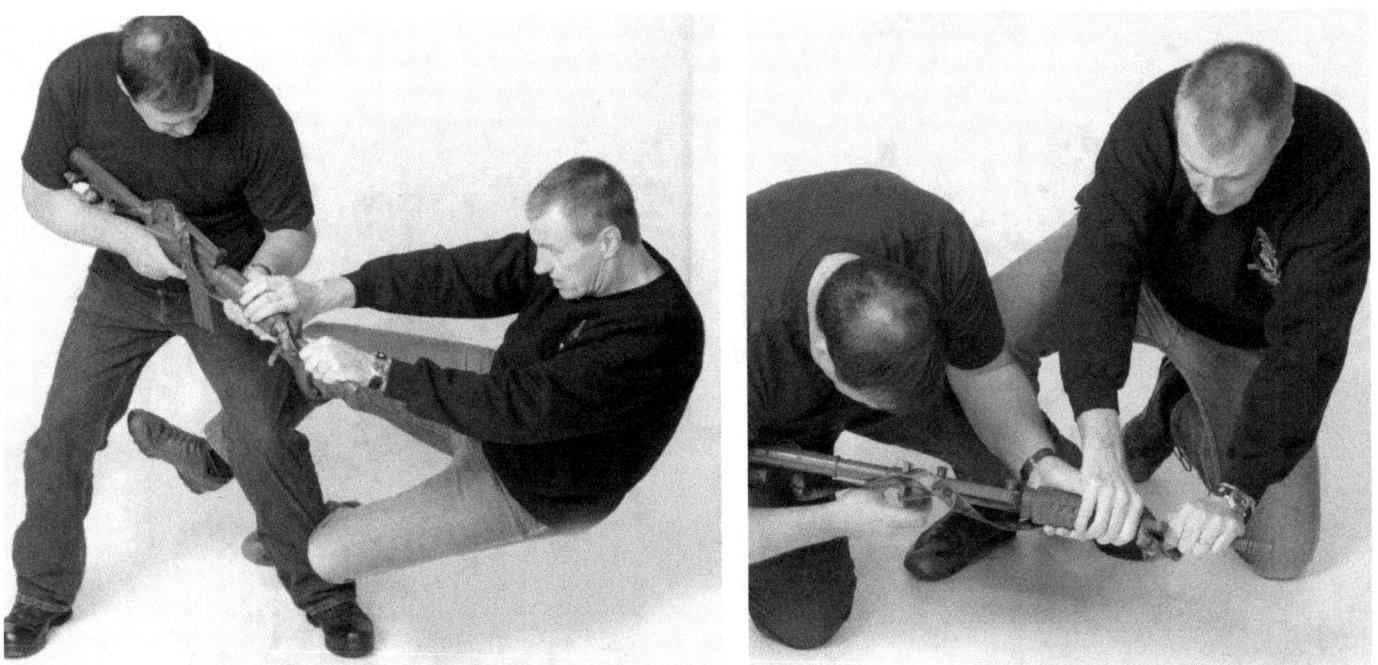

With a solid grip on the barrel, you must propel yourself against the enemy's legs and take him off balance.

Control the weapon, and bash his head and neck.

Long Gun Disarm Scenario 10: The Snap Shooter

You've walked right into an ambush, or maybe a robbery. The enemy raises and points his weapon at you to shoot. You clear the barrel, draw your pistol and fire.

The capture.

Barrel clearance.

The draw.

You can choose to shoot low into the pelvis, but his brain may still function. You may pass your forearm and shoot at his neck, face and head and shut down more bodily functions. Either way, you must hang onto the weapon and ride the body and gun down, to keep that barrel off of you in case he fires the weapon as he dies.

Long Gun Disarm Scenario 11: Countering the Standing Mugger

Many criminals, enemy soldiers and terrorists take a domineering position. They straddle their victim, brandishing their weapon and issue all kinds of verbal orders or commands. In this scenario, we look at one quick escape idea, a push and pull on the lower legs of the attacker.

Hook and pull the legs!

**Hook the foot.
Push the foot and shin.**

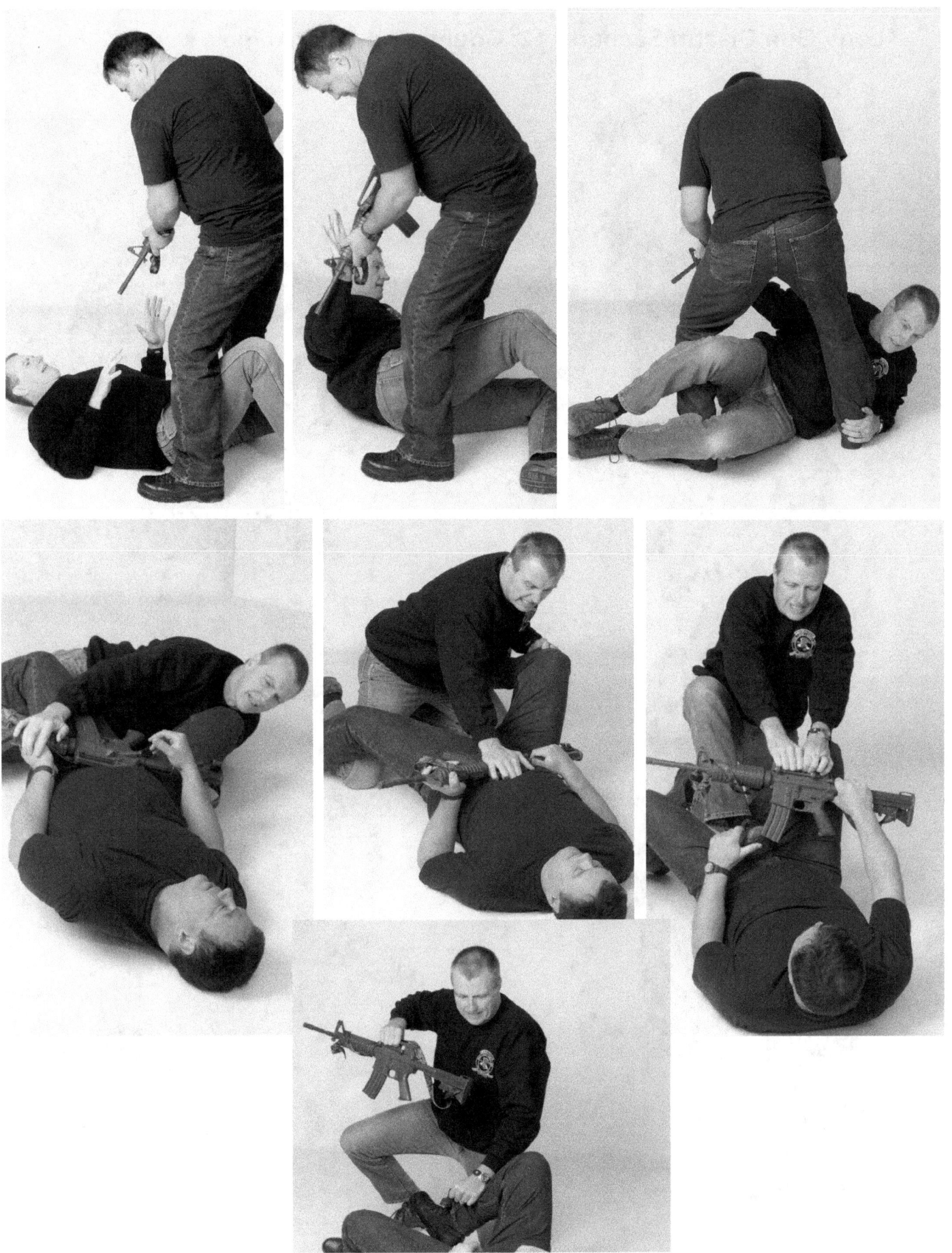

Page 189 - W. Hock Hochheim's Training Mission Five

Long Gun Disarm Scenario 12: Countering DMS Angle 1

The opponent executes a high right butt stroke.

You must dodge the butt stroke. Try and pass the stroke, if you can, since it will be delivered with momentum and power.

Two strikes, one to body and one to the head. Then rip the long gun from the stunned man.

Long Gun Disarm Scenario 13: Countering the DMS Angle 2 Attack

Evade the barrel (or bayonet) attack. Dodge the barrel and use footwork.

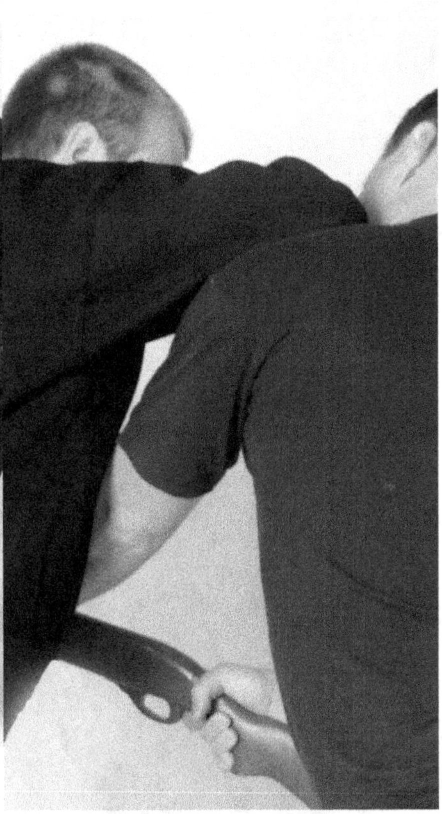

Dodge and pass with your hands if you can. Multiple strikes. Disarm.

Long Gun Disarming Scenario 14: Countering the DMS Angle 3

It is a medium-high impact attack from his right to your left, usually from the stock, since over 90 percent of the population is right handed.

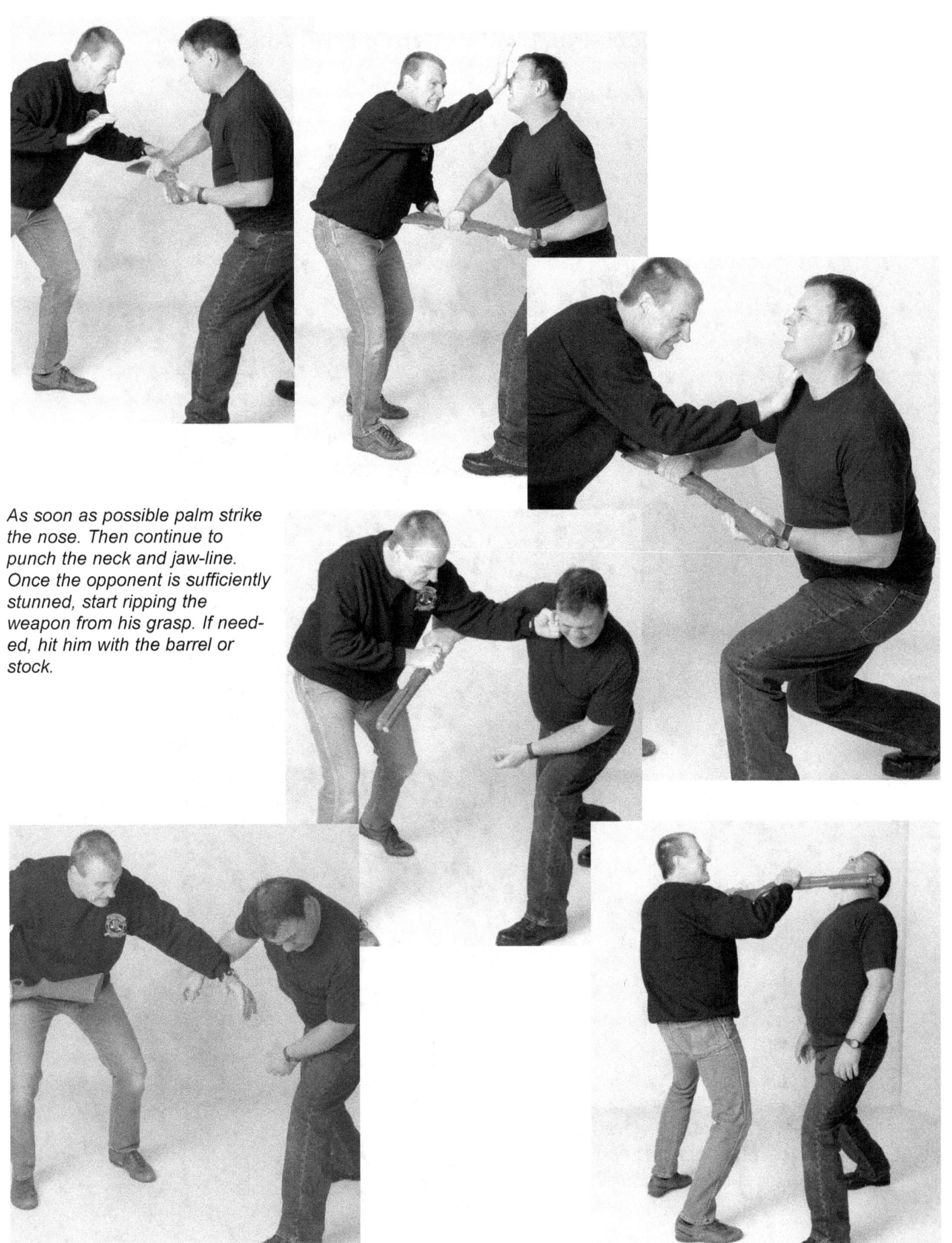

As soon as possible palm strike the nose. Then continue to punch the neck and jaw-line. Once the opponent is sufficiently stunned, start ripping the weapon from his grasp. If needed, hit him with the barrel or stock.

Long Gun Disarm Scenario 15: Countering DMS Angle 4

It is a medium-high impact attack from his left to your right, either from the barrel, or a bayonet slash.

You'll need body dodging and footwork to avoid the worst and most painful point of impact. Stay clear of the barrel.

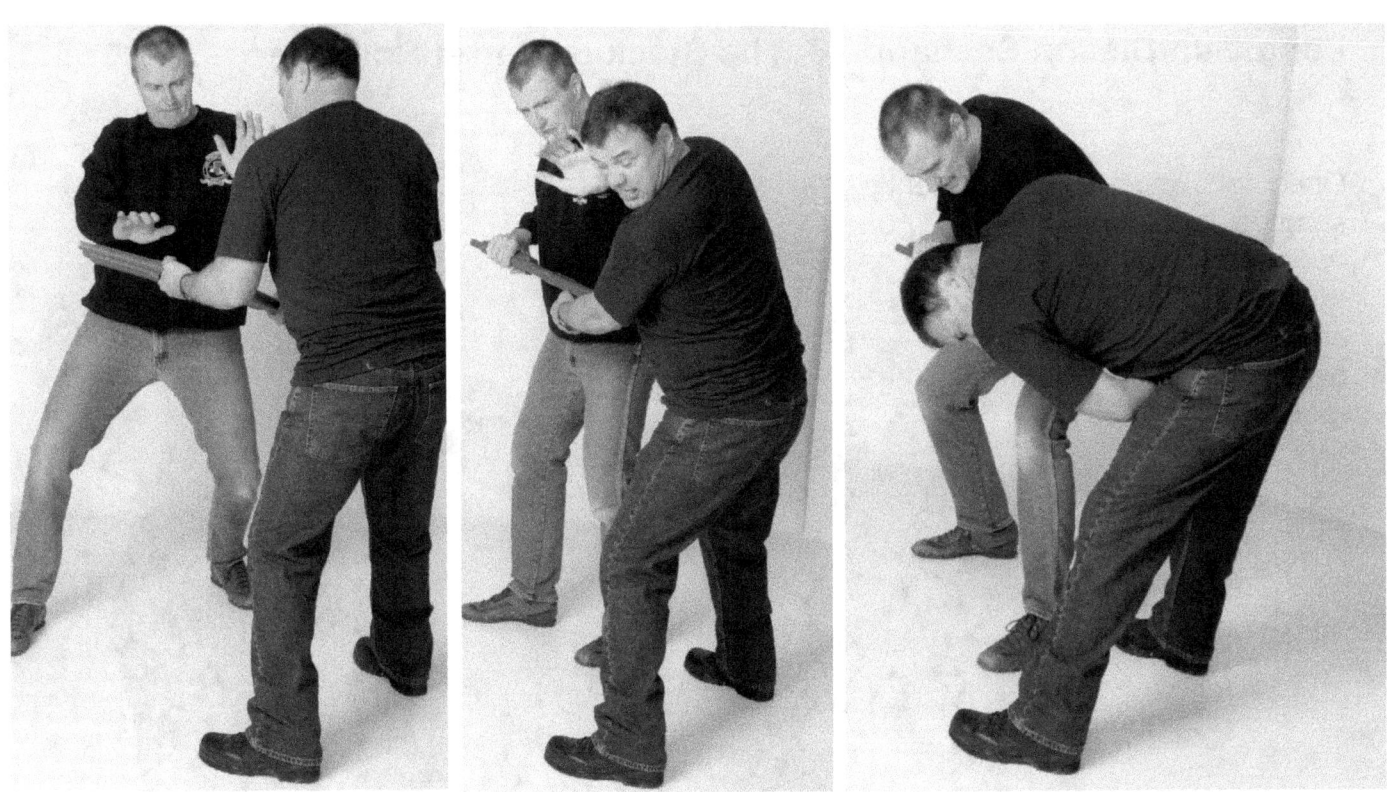
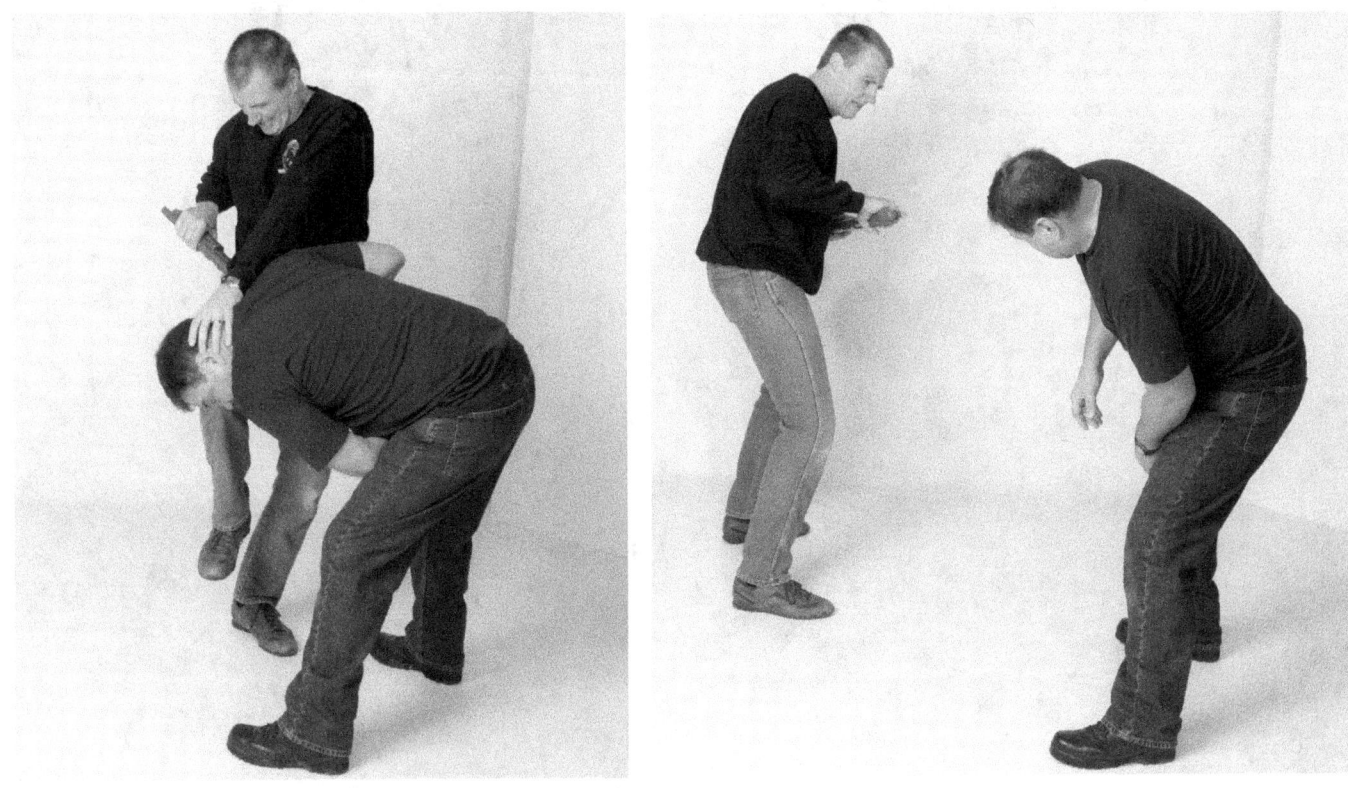

Long Gun Disarm Scenario 16: The Stock or Barrel Roll-Over

Once your opponent is sufficiently stunned, you may rip the long gun out of his grasp by taking hold of the barrel and/or stock and rolling one end over the other in a rowing motion.

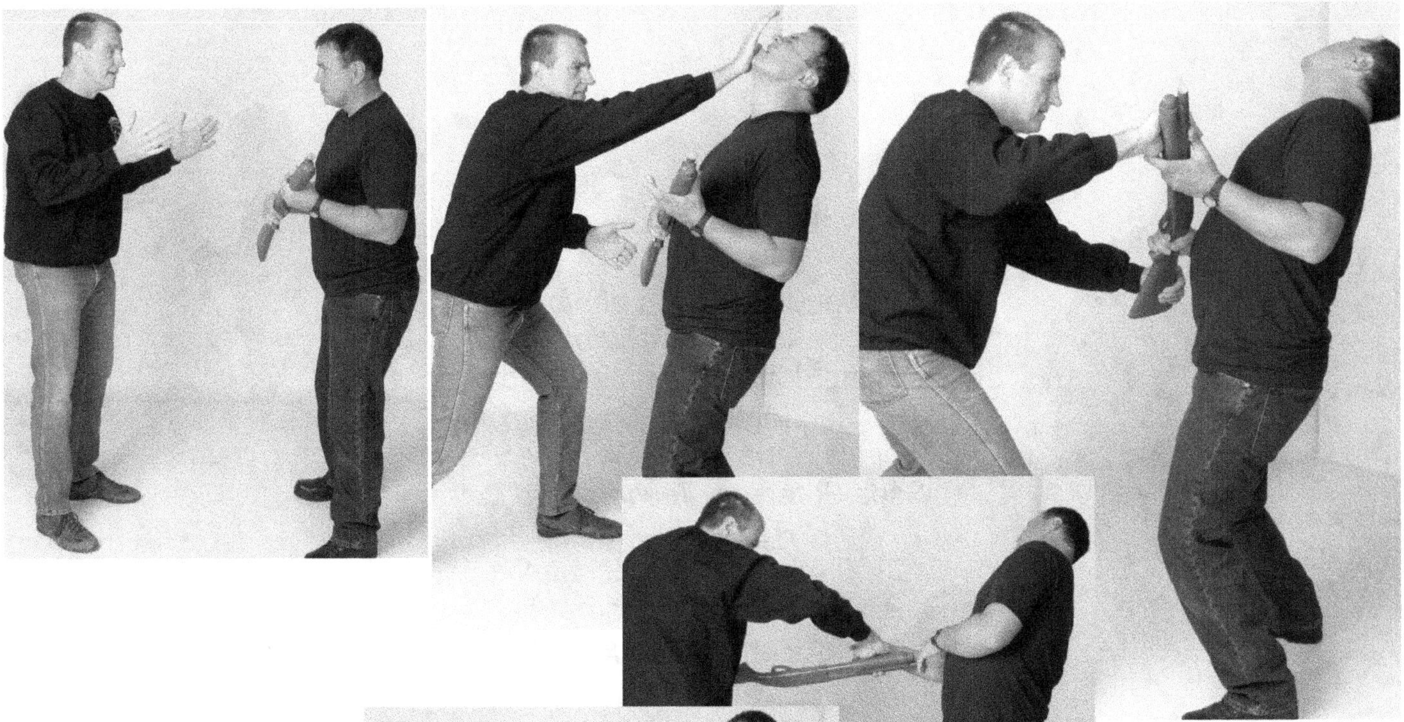

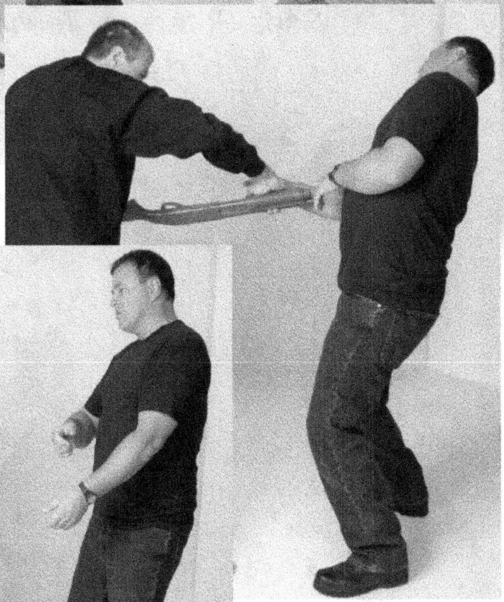

Ripping the weapon in a rolling manner to disarm the opponent.

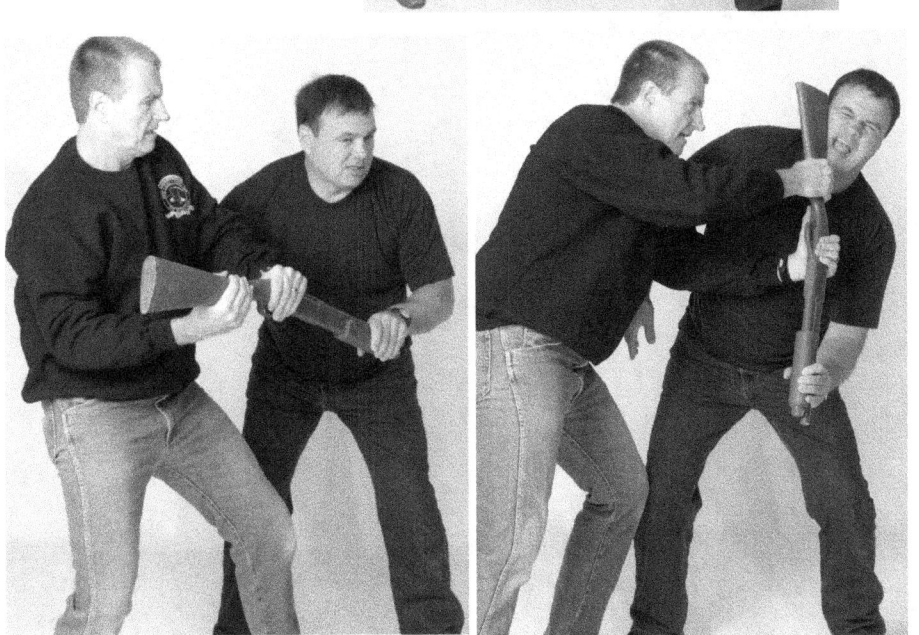

You will also see there are opportunities to roll the stock or the barrel into a strike, while in the process of attempting the disarm.

Take note of how you may use the free end of the weapon to smash the enemy, even while he holds the other end.

With Much Thanks To:

Jane Eden, *Editor and photographer*
Janne Hakkinen, *Stuntman*
Jeff "Rawhide" Laun, *Stuntman, photo facility*
Stefan Mattsson, *Stuntman*
Thomas Pentzer, Associate editor
Scott Peterson, *Technical editor*
Randy Roberson, *Stuntman*
Dr. Curtis Sheldon, *Stuntman*
Bill Whitworth, *Photographer*
Allan Zant, *Photographer*